斯沃特-迈耶斯事务所作品集

[美] 拉塞尔·亚伯拉罕 编　齐梦涵 译

30
PROJECTS

斯沃特-迈耶斯事务所作品集

[美] 拉塞尔·亚伯拉罕 编　齐梦涵 译

广西师范大学出版社
·桂林·

images Publishing

30
PROJECTS

图书在版编目(CIP)数据

斯沃特-迈耶斯事务所作品集/(美)拉塞尔·亚伯拉罕编;齐梦涵译.—桂林:广西师范大学出版社,2017.4
ISBN 978-7-5495-9538-9

Ⅰ.①斯… Ⅱ.①拉… ②齐… Ⅲ.①建筑设计-作品集-美国-现代 Ⅳ.①TU206

中国版本图书馆CIP数据核字(2017)第031125号

出 品 人:刘广汉
责任编辑:肖 莉 齐梦涵
版式设计:吴 迪
广西师范大学出版社出版发行
（广西桂林市中华路22号　　邮政编码:541001）
（网址:http://www.bbtpress.com）
出版人:张艺兵
全国新华书店经销
销售热线:021-31260822-882/883
恒美印务(广州)有限公司印刷
(广州市南沙区环市大道南路334号　邮政编码:511458)
开本:787mm×1 092mm　　1/12
印张:$21\frac{1}{3}$　　　　字数:30千字
2017年4月第1版　　2017年4月第1次印刷
定价:256.00元
————————————————
如发现印装质量问题,影响阅读,请与印刷单位联系调换。

目录

6   前言：斯沃特-迈耶斯事务所，一段与现代主义的恋情

住宅

12   住宅的开端

18   **01** 奥兹住宅
30   **02** 辛巴达溪住宅
42   **03** 辛帕蒂科组合式预制住宅
50   **04** ARA住宅
60   **05** 斯坦别墅
70   **06** 维达拉基斯住宅
84   **07** 回顾葡萄园住宅
98   **08** 拉希德住宅
106 **09** 克诺尔住宅
118 **10** 成-布赖尔住宅
128 **11** 门斯特-迈尔斯住宅
136 **12** 成-雷因格纳姆住宅
144 **13** 莫拉地产

在建住宅

150 **14** 印度北部住宅
154 **15** 圣地亚哥住宅
158 **16** 太阳神之屋
162 **17** 阿玛拉住宅
166 **18** 私语之石住宅
170 **19** 入江住宅
174 **20** 欧西欧里住宅
178 **21** 西风住宅
182 **22** 查隆路住宅
186 **23** 李住宅
190 **24** 河畔住宅
194 **25** 黑点住宅
198 **26** 凯路亚海滩别墅
202 **27** 萨那住宅

商业、教育与机构项目

208 从地点与客户需求中产生的建筑

214 **28** 棕榈泉动物收容所
222 **29** 百老汇网球中心
230 **30** 安嫩伯格项目

238 项目版权信息
240 当前团队
242 人物介绍
244 部分年表
248 出版物
252 图片版权信息
254 致谢
255 索引

前言

斯沃特-迈耶斯事务所，一段与现代主义的恋情

拉塞尔·亚伯拉罕 (Russell Abraham)

建筑也许是一种最能引起人们强烈感官体验的艺术形式，因为它所引起的不是针对某一特定感觉的直接刺激。这里面没有优美的弦乐与在舞台上翩翩起舞的天使般的舞者，也没有沁透缤纷色彩的画布和令人陶醉的样式，它把自己包裹起来，使人产生一种空间感或场所感。建筑自身与土地相连，构成一个人们可以在其中生活的空间。而房屋如果超出了其使用功能并带有美学意义，便成了建筑。在这个独特的景观、结构和美学的网络中，斯沃特-迈耶斯事务所发展起来。在过去的二十几年中，斯沃特-迈耶斯事务所成为加利福尼亚州创新设计领域里代表21世纪现代主义和高品位的灯塔。

加利福尼亚位于美洲大陆的西部边缘，挤向太平洋板块，构造力促成了当地多样的地形，形成起伏的山峦、土地肥沃的山谷和积雪盖顶的山峰等美丽的景观。这种崎岖的景观使这里在20世纪变成艺术和科学技术的大熔炉，催生出一种混合文化，而这种文化在后来逐渐渗透到整个文明世界。20世纪20到30年代，这里成为来自欧洲的政治和文化难民们聚集的地方，伴随他们一同前来的还有这些人的审美观念。鲁道夫·辛德勒 (Rudolf Schindler) 和理查德·诺伊特拉 (Richard Neutra) 来到加利福尼亚，与弗兰克·劳埃德·赖特 (Frank Lloyd Wright) 一起在洛杉矶工作；埃瑞许·孟德尔松 (Erich Mendelsohn) 逃离纳粹德国，最终留在伯克利；阿尔瓦·阿尔托 (Alvar Aalto) 曾于20世纪50年代短暂地逗留此地。他们带来了包豪斯理念，将其与赖特的田园派现代主义审美相结合，创造出一种部分工艺美术风格、部分包豪斯风格、部分日式风格的混合加州现代主义风格。第二次世界大战结束之后，加州凭借自身光鲜亮丽的产业结构与温和的气候，吸引各地人们和各种文化汇聚于此。20世纪30到60年代之间，洛杉矶市的人口翻了一倍，洛杉矶县人口增长得更多。这里的文化包含了上一代包豪斯风格建筑师们所拥护的现代主义美学，也正是在这种具有前瞻性思维的文化中，罗伯特·斯沃特 (Robert Swatt) 第一次认识了设计与美。他在洛杉矶西部长大，赶上了城市当中建筑物急剧增多的时代，这些建筑物大多是20世纪中叶时期的现代主义风格。他所就读的初中的教学楼是由理查德·诺伊特拉设计的。当时还在学校里读书生活的少年斯沃特切身体验了充满日光的明亮教室和周围优美的庭院，这给他留下了不可磨灭的印象。

斯沃特在加州大学伯克利分校环境设计学院度过了他的大学时光。在这里，他结识了一些美国建筑界的重要人物，并向他们学习。他很幸运地师从全国知名且极具影响力的建筑师威廉·特恩布尔（William Turnbull）和唐纳德·奥尔森（Donald Olsen）。不过，斯沃特与洛杉矶著名建筑师雷·卡佩（Ray Kappe）建立的师生关系对他来说则更加重要。卡佩是南加利福尼亚地区设计界的重要人物，他不仅是一名优秀的建筑师，更是南加利福尼亚州建筑学院的创办者，该学院后来成为全美国最主要的建筑学院之一。

斯沃特的事业起步于旧金山的霍华德·弗里德曼事务所（Howard Friedman），他后来搬到洛杉矶，加入格鲁恩公司（Gruen Associates）并与西萨·佩里（Cesar Pelli）共事。在那里，斯沃特完全认识到了自己创造美的欲望。在洛杉矶待了两年之后，斯沃特于28岁时回到南加州并创办了自己的公司。公司的第一个项目是一个投机性的低成本住房——阿米托一号（Amito I）——该项目荣获日落 - 美国建筑师协会奖（Sunset-AIA award），为他的新公司打响了知名度。阿米托一号是一个采用了解析几何的项目，其选址位于在一个极难处理的山坡上，只有最勇敢的人才会选择在那里建房。斯沃特有幸在20世纪80年代获得李维斯公司（Levi Strauss Company）这家客户，源源不断的工业与商业项目随之而来。这家客户使斯沃特和他年轻的公司不再总结过去，有些人将之称为误入歧途，也正因为这样，他们原本的美国建筑品位逐渐向后现代主义转变。斯沃特在这期间对多种设计形式进行了尝试，但是对哪种都不甚满意。他的心还是在现代主义上，他渴望成为这种设计风格的火炬传递手。

1995年，斯沃特和他的妻子克里斯蒂娜·波夫莱特（Cristina Poblete）在旧金山15英里以东的郊区购买了一块土地，为他们的家庭修建了一处住宅。这个占地4000平方英尺的房屋坐落于一个靠近小溪和当地公园的斜坡上。和诺伊特拉设计的初中校园极为相似，这座房屋也充满了阳光，南侧朝向一个上坡的院子，北侧有一个宽敞的平台。所有的公共空间均与一条装有天窗的走廊相连，这条走廊则通往一个周围种满了橄榄树的优美庭院，整座房屋的立体设计削弱了反复出现的水平元素。这座住宅当时获得多个主流设计奖项，重新确立了斯沃特作为加利福尼亚州与美国西部地区现代主义复兴运动的中流砥柱的地位。

自20世纪90年代起，该公司的主要业务开始趋向于为整个西部地区设计优雅的独户住宅。在这期间，公司完成了大量设计项目，其中大多数是住宅，另外还有超过25个正在进行中的项目。斯沃特完全掌握了现代主义的建筑语言，并把自己的独到之处融入其作品之中。斯沃特的作品从辛德勒、奥尔森和欧文·吉尔 (Irving Gill) 的作品中获得灵感，以自身独到的具有21世纪色彩的表达形式，展现出20世纪中叶的现代主义风格。无论是裸露的钢柱、暴露结构的开放式天花板元素，亦或是20英尺高的混凝土承重墙，斯沃特总是把结构当做设计元素的一部分。暴露这些结构，将其当做设计元素来使用，是他设计的一个特点。

大部分斯沃特设计的住宅都位于加利福尼亚州，那里气候温和，一年中有九个月的时间适合户外活动。因此，整合室内外空间就成了他需要优先考虑的事情。他的大多数作品都包含大面积的玻璃墙，有些墙可以卷进挂在墙上的外罩中，使大面积室内空间可以瞬间转化为户外露台。他常采用连续的地板材料、不可见的外部排水系统，以及最小化的导轨，使这种转化变得流畅自然。

斯沃特坚持"从内而外"地设计他的住宅项目，并把外部形态作为对内部空间的一种表现形式。这不是不重视建筑外形，而是希望每一座房屋都能展示出各自的外部表达和内部功能之间的相互作用。用优雅的硬木墙壁和屋檐来搭配粗糙的白色灰泥，以便软化其硬朗的视觉效果。将外部形态向前伸出或向后凹入，来打破枯燥的水平面，使其产生更美观、更令人愉悦的光影效果。使用水平的遮阳棚，一方面减轻了正午时加利福尼亚的暴晒，一方面又提供了令人赏心悦目的光影线。其结果就是一座看起来像是于20世纪50年代设计的、却又完全属于21世纪的建筑。斯沃特将其称为"穿越时间的美，抓住现代精神的产物"。

2009年，乔治·迈耶斯 (George Miers) 公司与斯沃特公司合并成为斯沃特-迈耶斯事务所，虽然迈耶斯公司的主营业务是公共和机构项目，斯沃特公司的主营业务是住宅项目，但是这却是一次成功的公司合并。其中一个原因是双方拥有一致的设计理念——两家公司都创办和发展于20世纪中叶现代主义盛行的时期，他们都不断坚持其最基本的理念，也是21世纪新趋势的积极倡导者。斯沃特和迈耶斯虽然依旧在其自身的专业知识范围内工作，但这次合并使两位建筑师有机会分享他们的知识和感受。

乔治·迈耶斯在旧金山的一个工薪阶层的社区长大。迈耶斯说他小时候对智力游戏很着迷，他那解决问题的能力和艺术家的性情把他带入了建筑行业。迈耶斯学习刻苦，运气也不错，还有很好的体育成绩，所以比较顺利地考进了位于圣路易斯的名校——华盛顿大学。华盛顿大学在20世纪60年代是美国一流的建筑院校之一。迈耶斯还能回想起该校的教职人员查尔斯·摩尔（Charles Moore）、里卡多·莱格雷塔（Ricardo Legoretta）和著名的瑞士建筑师道夫·施奈伯利(Dolf Schnebli)在墨西哥圣米格尔德阿连德领导的一个学生工作室。他在那里明白了有关设计的一个重要道理，那就是好的建筑不仅仅是一个好的解决方案。建筑师要理解建筑所服务的人群和文化，并把这种理解融入自己的作品中。

在华盛顿大学取得学位之后，迈耶斯有幸加入SOM建筑事务所旧金山分公司的设计团队，该团队当时由传奇人物查尔斯·贝塞特（Charles Bassett）领导。迈耶斯说由贝塞特集合起来的团队吸纳了当时的一些最有创造力的年轻建筑师，在那里工作让他既害怕又振奋。后来他被赫伯特·麦克劳克林（Herbert McLaughlin）挖到后来成为北加州和美国西北部地区重要区域性设计公司之一的KMD事务所。在那里，迈耶斯领导了一个设计工作室，承接各种商业和机构的项目。迈耶斯在1982年离开KMD事务所，创办了自己的公司，主要为旧金山不断扩大的东部地区设计商务办公楼，这些工作又慢慢被包括社区活动中心、公共安全设施和动物护理设施等公共机构的委托所取代。他的公司曾获得多个奖项，他也因其在动物护理设施中所进行的革命性工作，成为该领域家喻户晓的杰出建筑师。这家公司很快就收到了来自全美国和加拿大的电话，请他为他们的动物护理机构设计具有创意的作品。

斯沃特-迈耶斯事务所发展出一种独特而易于辨认的风格，在加利福尼亚州20世纪中叶现代主义风格的基础之上融合进温馨的天然材料制品，形成一种会让赖特（Wright）和辛德勒感到骄傲的风格，它逐渐在世界范围内获得认可。斯沃特和迈耶斯都认为他们是从内而外地设计建筑。事实也确实如此，身处他们设计的房屋内部的确是一种感官上的享受。金属和厚玻璃板等鲜明的制造材料被拿来与更加质朴天然的材料搭配使用，以创造出一种平衡的张力。就地浇铸的混凝土墙壁搭配红木楼梯和由顶及地的玻璃窗。建筑内部各个平面微妙地变换，楼梯在几乎令人眼花缭乱的时尚空间中悬浮，玻璃墙被卷动移开，将内部空间与广阔的天井、泳池和花园相连。内部与外部的空间以一种质朴的方式无缝对接。斯沃特和迈耶斯幸运地拥有来自世界各地愿意在市区和郊区选择风景优美的地点修建自己的住宅的人们作为客户。斯沃特-迈耶斯事务所设计出的房屋不是仅为生活而存在的机器，更是淳朴却又灵活的载体，既适合每个地点，又能和谐地融入其周围环境之中。

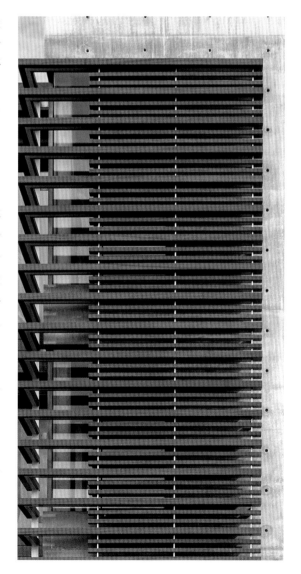

Above: Vidalakis House louver detail

住宅

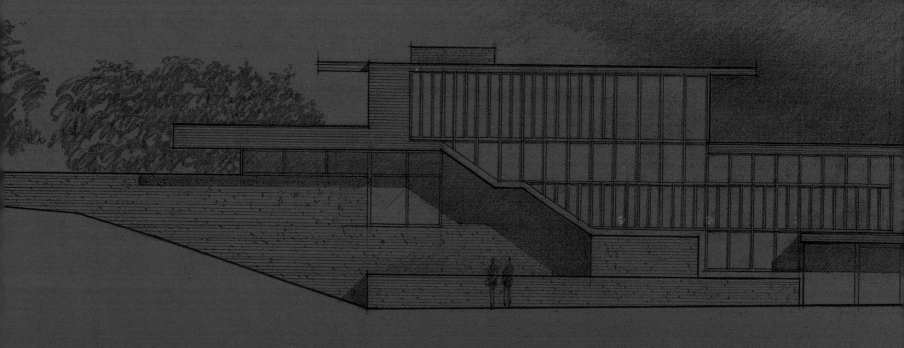

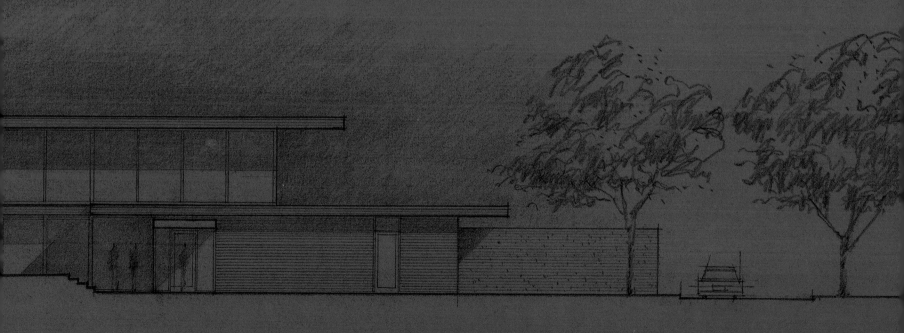

HOUSES

住宅的开端
RESIDENTIAL BEGINNINGS
Robert Swatt, FAIA

I have always loved designing houses.

There is something about the scale of a house, even a large house, where the essence of a design can be expressed with just a few strong architectural moves, where the story can be told with just a few words. Even the designs of complex houses on unique and challenging sites can, with patience and perseverance, be reduced to the simple expression of one or two basic concepts. In a time where many architects choose to take a relatively simple design problem and make it look as complex as possible, we choose to take the opposite approach and reduce every project to its essence. We aim for quiet simplicity in lieu of over-exuberant complexity, and we have found that the more we focus on the timeless principles of modern architecture, the more timeless the architecture in reality becomes.

So, what are the key timeless principles of modern architecture? First, since architecture is about life, we approach the design from the inside out, not the other way around. As a reflection of our informal West Coast family lifestyle, open planning—both vertical and horizontal—provides connectivity and spatial excitement to the interiors of our homes. In response to the wide range of site typologies, from expansive vineyard sites in the California wine country to steep hillsides in the North Bay and East Bay communities and semi-rural parcels with mature landscaping in Silicon Valley, we knit our buildings to their sites in such a way that they appear almost inevitable—with site and architecture united in a single, expressive composition. Connecting inside and outside, both visually and functionally, with generous areas of fixed and operable glazing, ensures that our buildings take full advantage of our beautiful landscapes and temperate climate. These key principles—designing from the inside out, open planning, knitting the building to the site, and blurring the boundaries between inside and outside—are the constant threads that connect all of our work.

In the early years of my architectural practice—the ten years beginning in 1975—I had the opportunity to test out various design approaches through the medium of house design. The early houses were one-of-a-kind investigations of design processes and expressions, with one constant: they were all built in the Berkeley Hills, on rugged, steep sites that in earlier times were considered unbuildable. The way these early homes related to the land, often stepping up or down the hillside in concert with the topography, became the overarching theme and expression of designs of this period.

A decade of commercial and corporate work followed these early residential explorations, culminating with The Icehouse project, a landmark transformation of San Francisco's largest pair of masonry buildings, originally built in 1918, into state-of-the-art headquarters space for Levi Strauss & Co. This was a complex project involving the insertion of 200,000 square feet of modern corporate interiors into historic timber-framed, brick-clad structures originally built for another purpose—to store ice for San Francisco's fishing industry. Consistent with our approach to residential work, the design reconciles the inherent complexities with a simple, unified expression in which the new elements beautifully coexist with the historic wood, steel, and brick elements. Nothing is hidden. The use of warm colors and natural materials adds to the sense of understated quality, in keeping with the Levi's motto "quality never goes out of style."

In 1995, I reconnected with my passion for residential design, starting with my own house. It had been four years since the devastating fire that destroyed 3,000 homes in the Berkeley and Oakland Hills, and having witnessed the insensitive and out-of-scale reconstruction of our neighborhood, my wife and I decided to move our family out of the area and build a new home on a semi-rural, undeveloped hillside in Lafayette, California, about 10 miles east of Berkeley. This was a watershed project for me, allowing for fresh investigations into what makes California architecture so unique. Terracing down a north-facing, oak-studded hillside, the design steps with the land, creating strong interior-exterior connections on three levels. Spatially, the home is open both horizontally and vertically, with few walls and doors, resulting in informal shared living spaces suited to our family lifestyle, and a common theme in West Coast Modern architecture. Large glass openings on all levels maximize views and access to the surrounding natural environment, blurring the boundary between inside and outside. With interlocking spaces and contrasting warm and cold materials, terrazzo floors and exposed glue-laminated Douglas Fir beams on the interior, and cement plaster and cedar siding on the exterior, the design is both dramatic and at the same time warm and inviting—contrasts that are hallmarks of our work. The architectural language is a combination of the crisp geometry of cement-plaster volumes, reminiscent of European Modernism, and vertical and horizontal rhythms more commonly associated with Japanese architecture—a sort of East-meets-West composition. Importantly, however, the design is based on living, on space, not form. It was designed from the inside out, not the other way around. The basic principles involving the relationship of building to site, open planning for informal living, and dissolving the boundary between inside and outside have become important concepts in all of the residential work that has followed.

Top: Swatt House, elevation
Bottom: Swatt House, living room and terrace

Six years after designing my own house, we were commissioned to design a small, very unique project in Saratoga, California, near Silicon Valley, on a gentle hillside under magnificent heritage oak trees. The project brief was simple enough; three individual tea houses for three distinct functions. One tea house for meditation, one tea house for overnight guests to sleep, and one tea house for work. This project led to an exploration of the tension between architecture and site, between anchoring and floating. Supported by cast-in-place concrete U-shaped tower elements taking all vertical and horizontal structural loads, light steel-framed glass boxes were designed to float above the ground like delicate Japanese lanterns. The crisp geometry and detailing heightens the contrast between architecture and land, between the theoretical and the organic. This tension between built and unbuilt has found its way into several new projects where major building elements hover above the land and sometimes appear to soar into space.

Top: "Steeping" tea house
Bottom left: "Visioning" tea house entrance
Bottom right: "Visioning" tea house interior view

In 2005, four years after designing the Tea Houses, we were asked to transform a tired and poorly built 1960's T1-11 plywood home into a new, expansive modern home that could take advantage of some of the best views in the San Francisco Bay Area. The Garay House in Tiburon, California, further explores the concepts of open planning, connections of inside and outside, and dissolving the boundaries between the two. With a minimal palette of vein-cut Turkish travertine and walnut casework on the interior, and cement plaster and Jerusalem st on the exterior, the composition is a quiet and elegant backdrop for enjoying the outdoors and spectacular views of two prominent Bay Area landmarks—San Francisco Bay and the Golden Gate Bridge. The simplicity of the building forms and materials yields a quiet composition, reduced to its essence—a low, transparent pavilion perched above the bay.

The houses featured in this book cover much of our residential design from 2006 to 2016. This period has seen our work expand into different regions of California, including Central Valley, the Wine Country, and Los Angeles; different states, including Hawaii and Colorado; and different countries, including Canada, India, and Spain. Today, our residential clients come from all over the world.

In the Oak Knoll, Helios, and Retrospect Vineyards houses in Northern California Wine Country, and in the Cheng-Brier house overlooking the bay in Tiburon, we continue to explore the dynamic contrast of architecture and land. The OZ, ARA, and Lee houses, all located on flat suburban sites, create their own surrounding environments through the integration of architectural and landscape design. Whispering Stones House is a delicate and ultra-sensitive modern insertion into a pristine woodland site in Healdsburg, California, and the Vidalakis House in Portola Valley is an almost elemental composition of wood, concrete, and glass in response to the owner's request for a design that is as much a work of art as the collection it houses.

In the mid-1970s when I was starting my practice, I taught architectural design at University of California, Berkeley. One of my colleagues, Sandy Hirshen, who later became head of the Department of Architecture, was a wonderful architect, friend, and long-time teacher at Berkeley. Midway through my first semester of teaching I casually mentioned that I had some doubts about the direction of my professional practice, but Sandy assured me that it takes patience and two years to know where you are headed in architecture. We had a similar conversation after two years of practice and Sandy said it really takes five years to figure out where you are headed in architecture; after five years he revised the number to ten years—and by that time I stopped counting.

In 1960, Le Corbusier wrote *Creation is a Patient Search*. After 40 years of practice, Corbusier's manifesto is finally beginning to sink in. As soon as you think you know something, there are always new challenges, new sites, and new clients with new requirements—all suggesting new design responses. That is one of the great things about the practice of architecture—every day seems like a new beginning.

Left: Garay House, dining area and master bedroom open to pool terrace
Right: View from lower terrace to master bedroom

01
OZ HOUSE

01

奥兹住宅
OZ HOUSE

ATHERTON, CALIFORNIA
DESIGN / COMPLETION
2006 / 2010

The first requirement the owners, a young couple with two children, expressed was that their new home should have a casual, barefoot feel, like a South Pacific vacation destination, with strong indoor-outdoor connections. The 2.8-acre site, with gentle slopes to the south and mature landscaping on all sides, was the perfect setting to create a home that would capture the ambience of a vacation retreat and fully engage the beautiful landscape for family living.

The design is based on a simple L-shaped plan with two wings. The east wing includes the kitchen and family room on the first floor, with children's bedrooms located on the second floor. The south wing consists of an office, media room, and guest suite on the first floor, with the master suite on the second floor. Connecting the two wings is a living/dining great room, fully glazed on the north and south sides. Under a low, cantilevered overhang, a solid mahogany pivoting entrance door opens to the dramatic great room, a beautiful two-story volume pierced by a floating glass bridge that runs east-west connecting the two wings on the second floor. With ceilings and two walls of Honduran mahogany, and two walls of floor-to-ceiling glass, this space recalls the indoor-outdoor lobbies of grand resorts in the South Pacific.

Materials have been selected to reinforce the architectural parti. The east and south wings are clad in white integral colored stucco as a counterpoint to the mahogany-clad central form of the great room. Deep cantilevered roof extensions, sheathed in mahogany boards at the central space and white stucco at the wings, reinforce the overall horizontal composition, visually dissolving the boundary between inside and outside.

The landscape has been designed to create two distinct environments. The north side includes a motor court, adjacent to an entry courtyard of rectangular stepping stones over a shallow reflecting pool. The south side of the house is accessed by glass doors from the living and dining areas, media room, kitchen, and family rooms, and has been designed for family living, with generous stepped terraces, lawn play areas, a barbecue patio and a swimming pool. A giant heritage oak tree, centered on the main terrace opposite the great room, has been preserved as a special focus viewed from the entrance and main living spaces. As the landscape steps with the natural grades towards the south, the formal aspects of the design begin to dissolve into a more naturalistic composition, contrasting with the orthogonal geometry of the architecture.

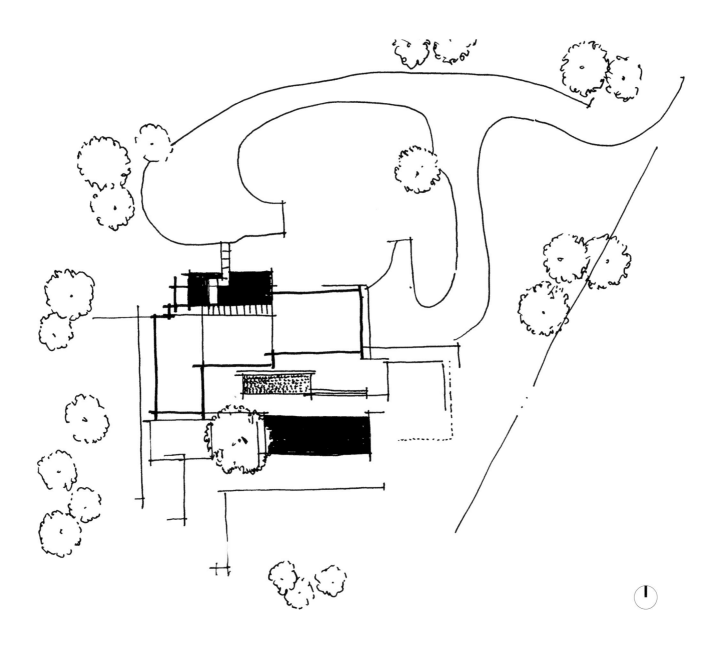

Previous pages: Entry approach
Above: Site plan
Following pages: Entrance elevation

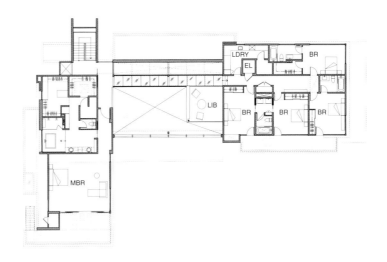
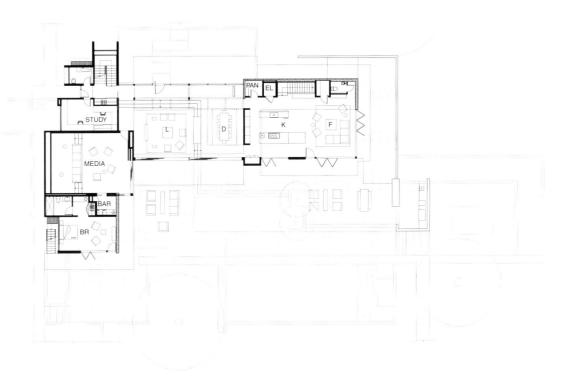

Left: Light enters through the skylight to illuminate the entry space
Above: Floor plans
Following pages: A large heritage valley oak tree is the focus of the south terrace

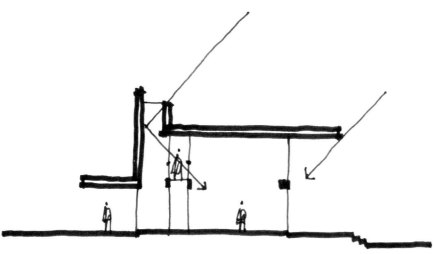

Opposite: Great room
Top: Glass walkway above great room connects the second-floor wings
Bottom: Section concept sketch

27

02 SINBAD CREEK RESIDENCE

02 辛巴达溪住宅
SINBAD CREEK RESIDENCE

SUNOL, CALIFORNIA
DESIGN / COMPLETION
2008 / 2010

The Sinbad Creek Residence started in 2008 when the owners—a couple, newly retired from the biotech field, and avid horseback riders—decided they wanted to be able to spend more time enjoying the outdoors. They desired a change from condominium living and purchased a bucolic 5.3-acre hillside lot in rural Sunol, with unique views in three directions: a beautiful upslope hillside with mature oaks and a giant walnut tree to the east, a canyon ridge to the west, and Mission Peak to the south. They imagined a modern residence that would embrace its beautiful setting, take advantage of the site's many vistas, and allow for multiple ways of enjoying the surrounding natural environment.

Accessed from a driveway at the northern edge of the property, the house is organized around a linear circulation spine that runs from north to south. Major living spaces are located diagonally from each other across the spine, creating transverse views throughout the house. Ceiling heights vary from a low, compressed entry to a two-story dining room, bringing additional spatial drama to the composition.

Alternating spaces enjoy hillside and canyon-ridge views, while the rooms at the end of the linear spine—the living room on the first floor and the second-floor master bedroom—enjoy beautiful views of distant Mission Peak. Generous patios and terraces for outdoor living and enjoying the vistas are located on the east and west side of the house.

The form of the house is at once both simple and strong, with a tall, stucco-clad vertical utility core anchoring the composition on the west (entry) side. Major living spaces with wall-to-wall glass flank the vertical core on both sides. Deep cantilevered roofs and terrace overhangs frame views and visually extend the interior spaces to the exterior. At the south end of the building, a double cantilever—the roof overhangs the second floor and the second floor overhangs the first floor—creates a dramatic visual thrust in the direction of Mission Peak.

Previous pages: West elevation
Above: East elevation

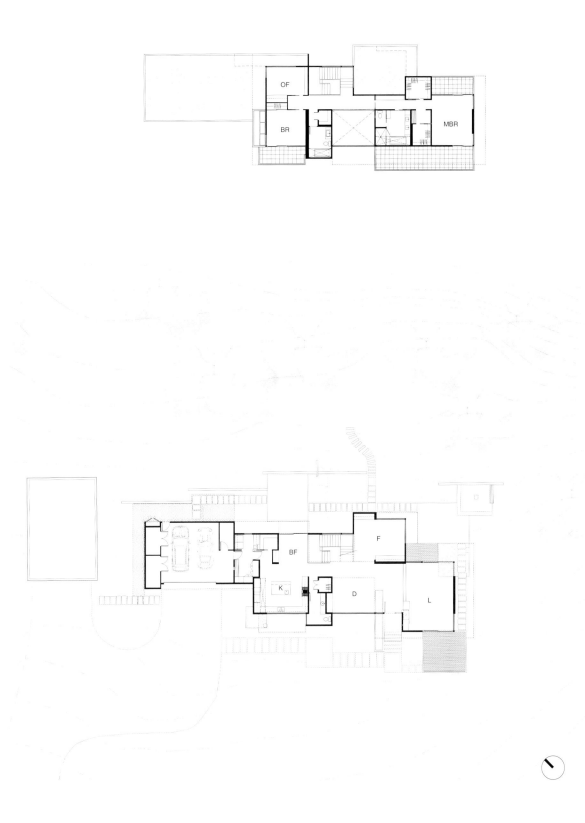

Left: House embraces its surrounding landscape
Above: Floor plans

Opposite: Staircase
Top: Entry at dusk
Bottom: West elevation sketch
Following pages: Family and living room

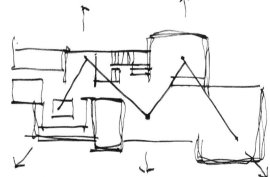

Opposite: View of dining room from entry
Top: Dining room and view of the surrounding landscape
Bottom: Plan diagram, transverse views

03
SIMPATICO PREFAB PROTOTYPE

03

辛帕蒂科组合式预制住宅
SIMPATICO PREFAB PROTOTYPE

EMERYVILLE, CALIFORNIA
DESIGN / COMPLETION
2007 / 2012

The partnership with Simpatico Homes represented an opportunity for Swatt | Miers to bring custom-quality architecture to a broader audience through the cost advantages of prefabrication. The Simpatico Prefab Prototype—Krubiner House—is located in Emeryville, California, just a few blocks from the Swatt | Miers's offices and was completed in January 2012.

Simpatico Homes are prefabricated structures specifically designed to take advantage of off-site manufacturing technologies. Built to the same code as site-built homes, the modular designs are constructed in a dedicated factory where quality control measures ensure a superior product. The homes are built in several smaller modules and are nearly 90 percent complete when transported to the site. On 'set day,' the modules transform an empty foundation into a full home structure by the end of the day.

The Simpatico Prefab Prototype was designed for a narrow, 30-foot-wide infill lot. At 2,407 square feet, this two-story home comprises two bedrooms, three bathrooms, a home office, and an extra study/sleeping loft. The home is designed for net-zero energy usage and a photovoltaic array on the roof is sized to produce sufficient power to meet 100 percent of electrical and heating demands. Varied ceiling heights, coupled with a unique plan that incorporates a secondary entrance/courtyard, gives this home a degree of customization not often associated with prefabricated housing.

In contrast to more typical prefab designs based on standard home plans, Simpatico is a proprietary modular design system that can be configured to suit the needs of the client and the demands of the site. Working with this 'kit of parts,' architects can create limitless unique designs within the system, and by doing so, the time and cost of design and fabrication are reduced.

The Simpatico Prefab Prototype was designed to meet the U.S. Green Building Council's LEED for Homes Platinum rating, positioning it among the most eco-friendly, sustainable, and healthy structures on the market.

Previous pages: Second floor exterior detail
Opposite top: Prefabricated units installed on site
Opposite bottom: Concept sketch
Above: Street view

Opposite: Entry
Top: Kitchen and dining area
Bottom: Living room

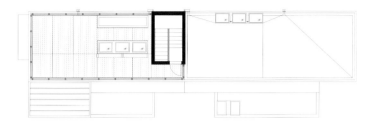

Opposite: Exterior view
Above: Plans

04 ARA住宅
ARA HOUSE

ATHERTON, CALIFORNIA
DESIGN / COMPLETION
2008 / 2012

This new family estate is located on a beautiful flag lot in Atherton, California. The setting provides total privacy from the neighboring houses, surrounded by mature oak, pine, cedar, walnut, and birch trees, and bordered by 20-foot-tall hedges along portions of the property lines. The earliest design strategy involved positioning the building on an angle, approximately 30 degrees, off of the orthogonal grid of the lot, creating diagonal vanishing vistas from each major space and increasing the sense of space surrounding the home. The trapezoidal-shaped exterior spaces on the north, east, and south sides of the property have been transformed into landscaped outdoor rooms for family living, with extensive private gardens, terraces, and swimming pool with adjacent changing room.

The house is approached from the west side via a long private driveway that skirts the edge of the property. A linear entry walkway, covered by a low, cantilevered roof overhang, leads to the main entry located at the intersection of two wings. The design is organized around an L-shaped plan that revolves around a three story, cast-in-place concrete stair core adjacent to the entry. A linear skylight on the west edge of the stair core brings natural light deep into the three-story stairwell, highlighting the texture of the board-formed concrete walls and providing visual interest to the entry. The stair core is joined by an additional concrete element—the living room fireplace—and together these concrete features visually anchor the architectural composition. The remainder of the house is clad with a single material—white integral colored stucco—as a quiet counterpoint to the cast-in-place concrete.

Functionally, the house is zoned with all of the public and family spaces on the first floor, and all private bedrooms, except for a first-floor guest suite, located on the second floor. A basement, covering approximately half of the building footprint, has been developed for a home gym and wine cellar.

Previous pages: East elevation
Above: Entry space

Left: Entry approach

Opposite: Staircase and second floor landing
Top: Light filters through skylight to highlight staircase and concrete wall
Bottom: Living room, with view to the dining room

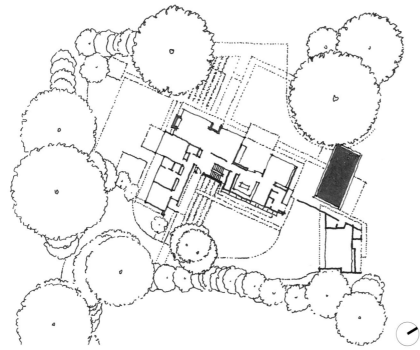

Opposite top: Doors open for indoor-outdoor seating
Opposite bottom: Site plan
Left: Shaded exterior seating

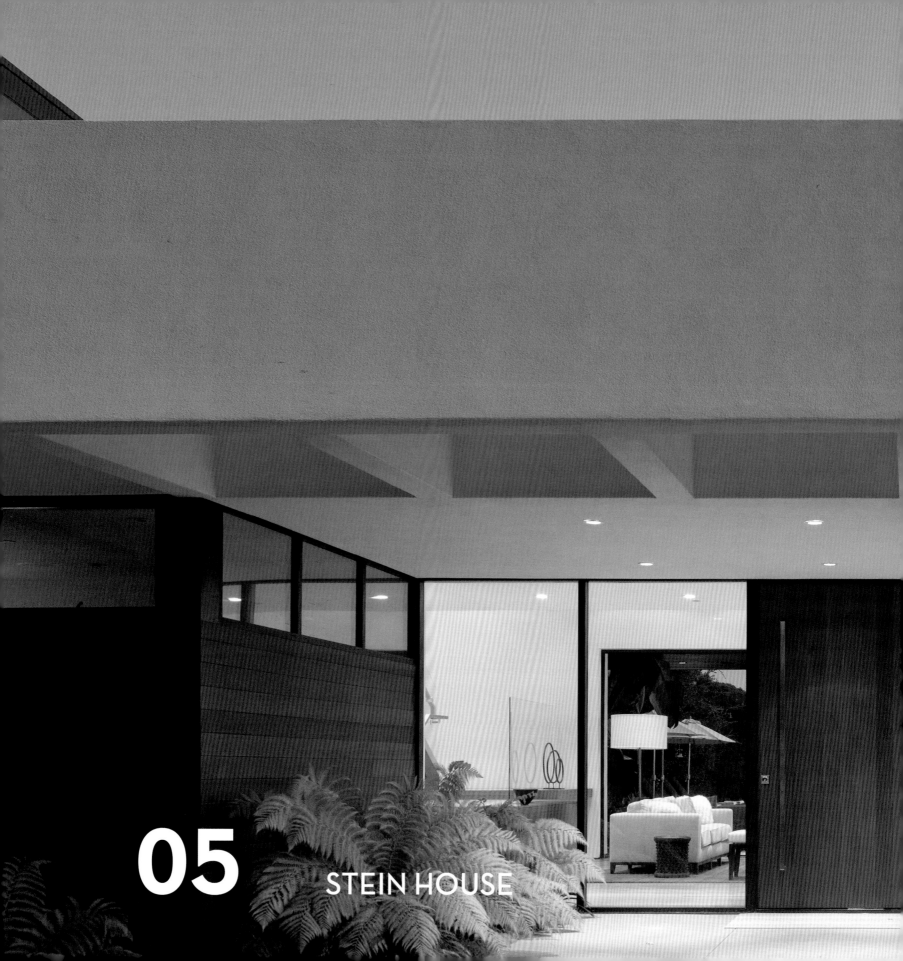

05
STEIN HOUSE

05 斯坦别墅
STEIN HOUSE

ORINDA, CALIFORNIA
DESIGN / COMPLETION
2008 / 2012

On a majestic ridge with sweeping views of bucolic hillsides and valleys, Stein House is an extreme makeover of a pre-existing one-story, tile-roofed ranch house. To reduce construction costs, the new home was constructed using the original foundations and much of the original footprint.

Due to the length of the footprint and the magnificent views to the west, the design strategy was to create a long path along the northern edge, culminating with an entry court centered on the east-west axis. The progression toward the entry is marked by the rhythm of a deep, open-welled trellis overhead. The entry splits the composition into two wings connected by a double-height living/dining space. Children's bedrooms are located in the east wing and connected by a floating bridge to the master suite in the west wing. The master suite, with its cantilevered terrace, enjoys unobstructed views of the Orinda valley and evening sunsets.

Landscape features include a large patio, pool, and fire pit on the south side of the house, a water feature marking the entry court on the north side, and a tennis court further north. Despite the constraints of using an existing foundation, the building feels entirely new, with a strong spatial organization that reorients all of the main spaces to take full advantage of the beautiful views from this hilltop setting.

Previous pages: Entrance elevation
Above: Northwest elevation
Following pages: Pool and terrace

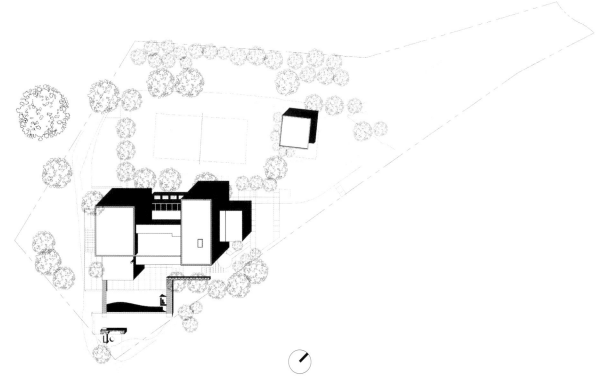

Opposite top: East elevation with pool and deck
Opposite bottom: Site plan
Right: Entry approach

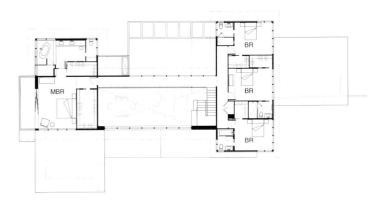

Left: Water feature and roof overhang at dusk
Above: Floor plans

06
VIDALAKIS HOUSE

06

维达拉基斯住宅
VIDALAKIS HOUSE

PORTOLA VALLEY, CALIFORNIA
DESIGN / COMPLETION
2009 / 2014

This distinctive residence was designed as a piece of inhabitable art for its owner, an avid collector of colorful and unique works. The design features three stepped-back stories linked by a grand mahogany staircase and a double-height glass pavilion overlooking the semi-rural landscape on three sides.

The client had contemplated building their dream house for years before finding the perfect site in Portola Valley, a semi-rural enclave south of San Francisco. The 3.8-acre site is a beautiful, gently sloping parcel, with mature oak, cedar, and pine trees sufficiently dense to provide privacy from neighboring houses, yet open enough to afford beautiful views of the surrounding hills as well as of distant San Francisco. The client's design requirements included open planning for interior spaces, zoning of functions to provide a clear separation between public and private spaces, maximizing views of San Francisco to the north, maintaining a strong relationship between building and landscape.

The resulting design is based on an L-shaped plan, anchored by a three-story, linear, cast-in-place concrete wall that projects into the landscape to frame outdoor spaces on both sides of the house. Main public functions are housed on the first floor, while the master suite and a home office are located on the floors above. The short leg includes a one-story home office and a detached garage and guest house. The plan organization creates two major outdoor areas: a private landscaped courtyard on the south side, and a more public viewing terrace on the north. Perpendicular to the house and topography, a 75-foot swimming pool, wood deck, and rectangular lawn define the western edge of the landscape with precise geometry.

A stand-alone art studio pavilion is located to the east of the main house on a direct axis with the front entrance. Placed on the east side of a cast-in-place concrete wall, this pavilion is virtually invisible as the concrete wall appears to simply be a background for art.

The architectural language is simple, almost elemental. Horizontal wood planes form the floors and roofs in counterpoint to the vertical cast-in-place concrete core, and are infilled with full-height glazing to maximize views. A rhythmic three-story window wall defines the circulation spine and stair and forms one edge of a private landscaped courtyard on the south side of the house. The main living space is two stories high and almost entirely transparent, with views of the valley in three directions.

Previous pages: South elevation
Above: Entry at dusk
Following pages: View of great room and entry

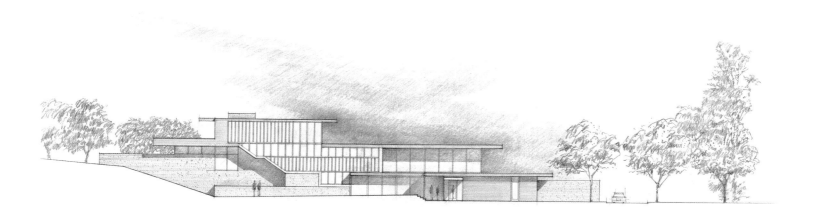

Opposite: North elevation detail
Top: Study model
Bottom: South elevation sketch
Following pages: Living room

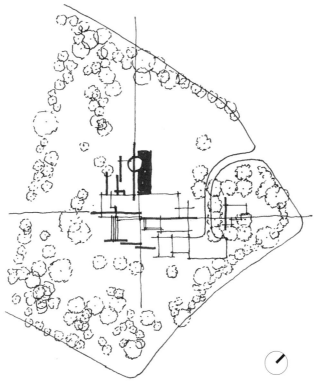

Opposite: Art studio pavilion
Top: Staircase along south facade
Bottom: Site plan

Opposite: Infinity-edge pool at sunset
Top: North elevation from far end of swimming pool
Bottom: Pool and sunshade structure

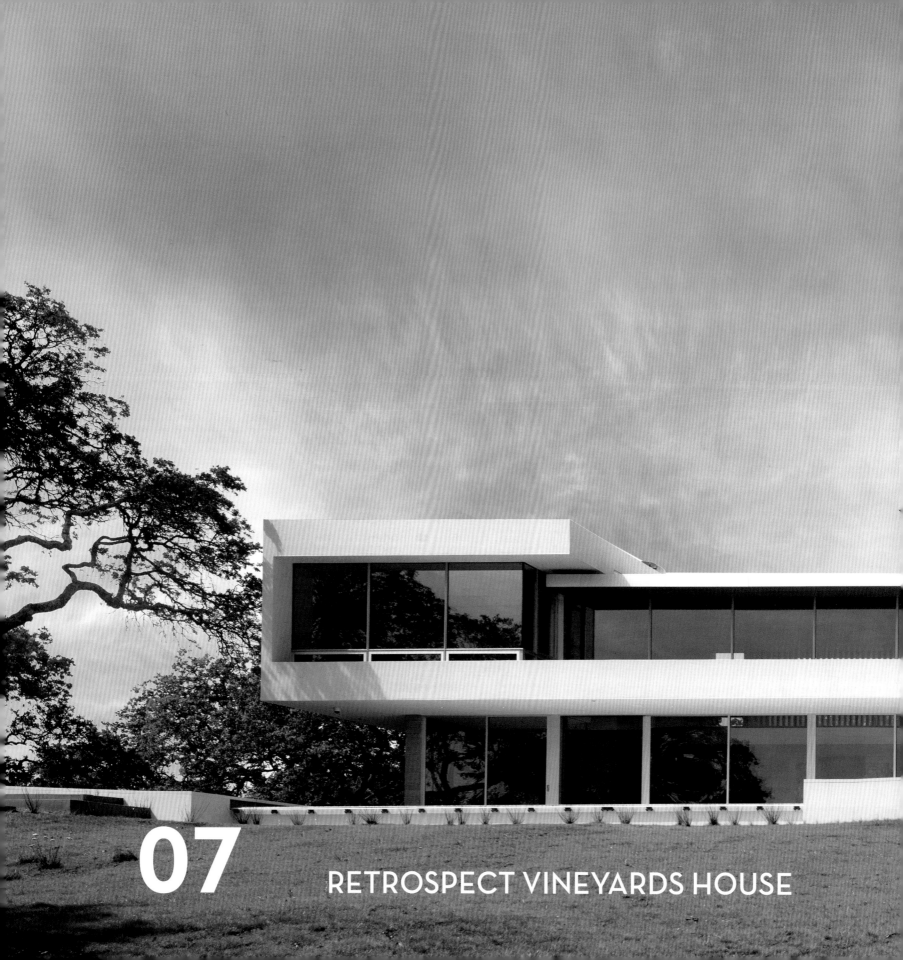

07 RETROSPECT VINEYARDS HOUSE

07

回顾葡萄园住宅
RETROSPECT VINEYARDS HOUSE

WINDSOR, CALIFORNIA
DESIGN / COMPLETION
2010 / 2014

Retrospect Vineyards has been producing pinot noir grapes and selling them to nearby wineries for fifteen years. When the new owners bought the property in 2010 they decided to have a greater involvement in the winemaking process and embarked on their journey to create a new modern home for their family, along with a working barn and staging area for their annual harvest.

Located on a gently sloping knoll surrounded by 20 acres of pinot noir vineyards, this new home has been designed to provide casual indoor-outdoor living spaces that take full advantage of its magnificent site.

Oriented on the site with business operations on one side and an expansive view of the surrounding vineyards on the other, the building was designed with two distinct façade treatments on either side of the narrow T-shaped plan: a relatively opaque entry court to the north provides privacy while admitting daylight through a delicate vertical wood-screen wall, while the south façade eschews solid walls in favor of double-height glazing.

Bedrooms are on the second floor, while public and shared areas are located below on the first floor. A double-height great room occupies the majority of the lower level with a 21-foot span of sliding glass doors seamlessly extending the living space to the outdoors. The great room, kitchen, and office all open up to a St. Tropez–limestone pool terrace.

The detached guest suite is envisioned as a glass box, just large enough to enclose a custom freestanding platform bed, projecting beyond the terrace edge. While the guest suite functions as a separate building, it remains connected to the main house by a low roof spanning the two structures. The guest space is screened from the main building while enjoying unobstructed 270-degree views of the landscape. The interstitial space between the structures is used as shaded lanai, with an outdoor kitchen and vineyard views on two sides.

Previous pages: North elevation from below
Above: Pool deck, facing west

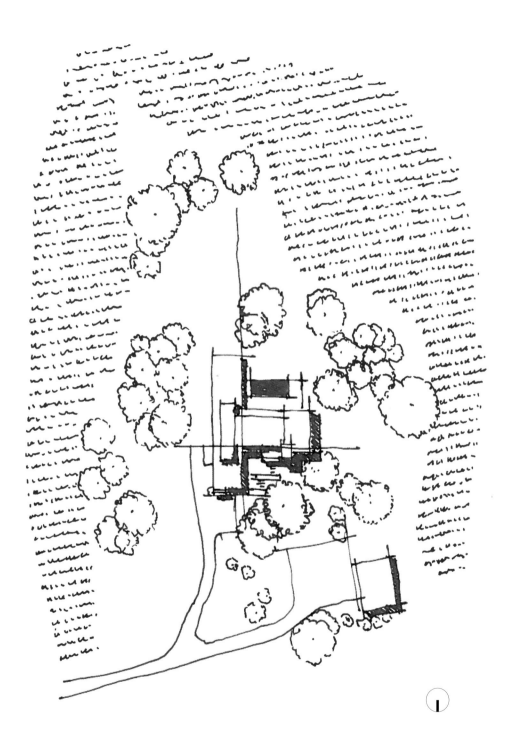

Opposite: North side of house
Above: Site plan

Left: View towards pool and guest room from great room

Above: Full-height sliding doors open the great room to the outdoors to create a seamless indoor-outdoor space
Opposite: Double-height great room, facing east

Opposite: Master bathroom
Above: Office with views of the vineyards

Opposite: View of guest suite from far end of pool
Top: Path along west edge of house
Bottom: Entry court

08
RASHID HOUSE

08

拉希德住宅
RASHID HOUSE

LOS ALTOS, CALIFORNIA
DESIGN / COMPLETION
2010 / 2012

Being of Indian heritage and having strong generational family ties, the Rashid family had to consider not only their needs and the needs of their two young children, but also the cultural requirements of their parents when designing their dream home. The resulting 6,800-square-foot house includes ample space for informal family living while taking into consideration the ancient Indian theory of Vaastu Shastra, an architectural doctrine that guides the location of spaces within a building based on the perceived effects of various natural forces.

The extended H-shaped plan divides the rectangular 1-acre site into two distinct outdoor zones: a public entry court to the east facing the street, and a private patio on the west that overlooks the pool, lawn, and gardens. The building façade reflects this distinction; the east side is predominantly opaque for privacy, while the west is largely transparent with large expanses of sliding glass doors that seamlessly connect to the outdoors. The sides of the H are defined by long stone-clad walls that run from east to west through the property, creating a sense of continuity and movement through the building while acting as a barrier between public and private functions. The warm-hued stone provides texture and contrast against the rich machiche siding and bright stucco exterior.

The kitchen, dining, and living rooms occupy the center of the H and are designed as one continuous volume. Despite their connection in plan, different qualities of space are created by the strategic use of different wall materials and ceiling heights. The kitchen and living room, both of which open up to the patio via pocketing glass doors, are differentiated by the living room's dramatic 20-foot ceiling, while the dining room takes on a more intimate feel thanks to the art that anchors it.

All of the family's bedrooms are located on the second floor, with the master and children's rooms separated from the guest room by a long skylit hallway overlooking the double-height living space. A second master suite was created on the ground floor to accommodate extended family visits. The owner's parents were closely involved in the design of this area, from the amenities of the attached art studio (the owner's mother is an artist) to the location of the suite at the north end of the house according to Vaastu principles.

Previous pages: Entry elevation
Above: West patio, with views inside

Opposite: Dining room from stairs
Above: Living room from dining room

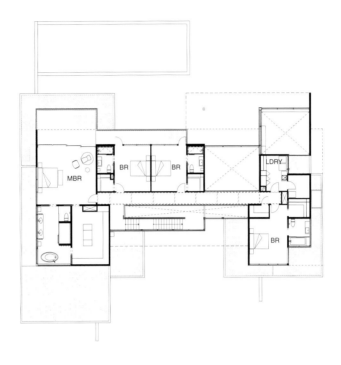

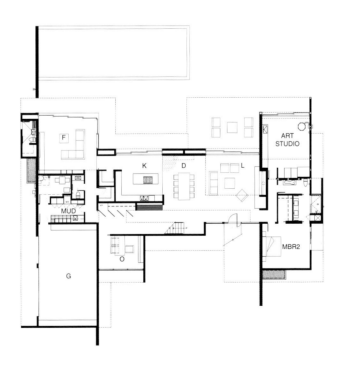

Opposite top: Kitchen from dining room
Opposite bottom: Master bathroom
Above: Floor plans

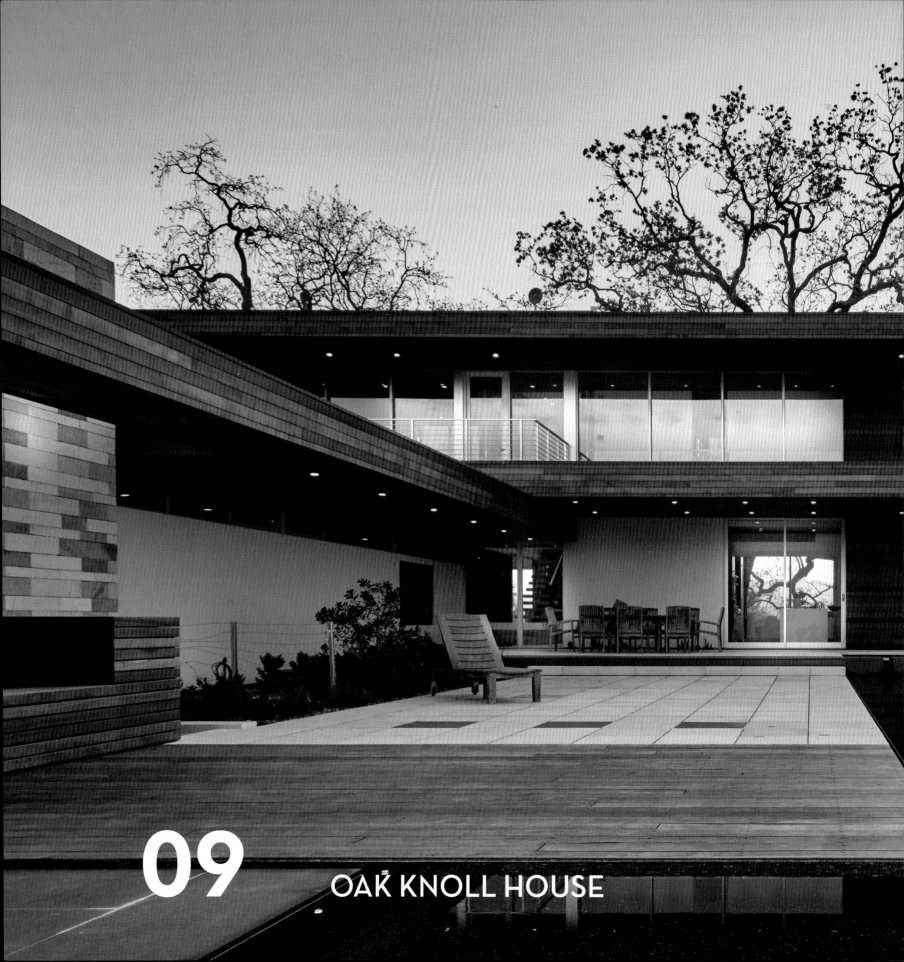

09 OAK KNOLL HOUSE

09 克诺尔住宅
OAK KNOLL HOUSE

NAPA VALLEY, CALIFORNIA
DESIGN / COMPLETION
2011 / 2016

The site had been on the owner's radar for many years and admired as they passed by on bicycle rides through the Napa Valley. With an oak studded knoll surrounded by 4.5 acres of cabernet sauvignon vineyards, the land has an almost archetypal quality—the perfect Napa Valley setting for a beautiful new modern home.

The dream started to become reality in 2011 when the property became available and the clients, a couple living in the Silicon Valley area, embarked on their journey to create their wine-country dream home.

The design brief was clear. The owners wanted a home that would respect and preserve the beautiful heritage oak trees that dotted the knoll; open up to the beautiful sunsets over Silverado Trail to the west; and have generous family outdoor living and entertaining areas oriented to the quiet, east side of the property with endless views of vineyards and rolling hills.

The plan has two wings. The north-south wing includes the two-story kitchen/dining/living space, as well as a guest suite on the first floor, and two home offices and master suite on the second floor. The master suite located at the north end of the wing is accessed by a dramatic glass bridge that traverses the double-height great room below. The great room and the master suite enjoy views of sunrises and sunsets in three directions. The east-west wing includes the garage and utility spaces and terminates at the east end with a private guest suite. A stone and glass-clad stairway tower, with a unique wine cellar embedded into the landing, forms the vertical anchor to the architectural composition from which the wings radiate.

One of the major themes of this design involves the relationship between architecture and nature. The master suite emphasizes this dialogue as it cantilevers over a stone plinth and soars into the canopy of one of the beautiful heritage trees, never quite touching the branches. As the branches twist and turn under and over the cantilevered terrace, it seems the home has always been there, in perfect harmony with its natural setting.

Previous pages: View of swimming pool, terrace, and east side of house
Above: West elevation at dusk

107

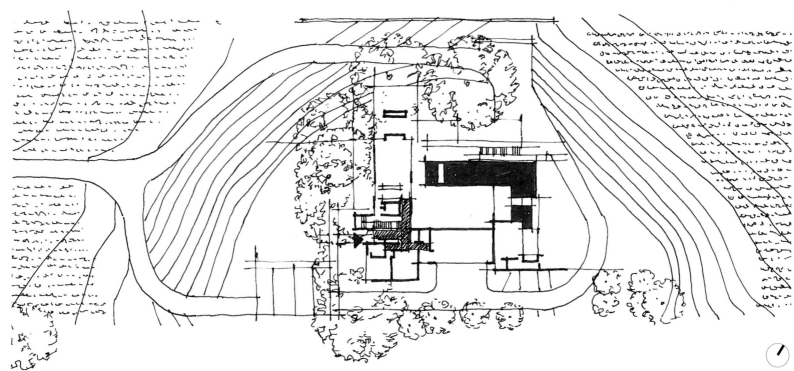

Top: West elevation
Bottom: Site plan
Opposite: Cantilevered terrace around master suite
Following pages: Elevation viewed from vineyards to the east

Opposite top: Kitchen, with view west
Opposite bottom: Stair landing at entrance
Above: Living room, with view north

Opposite: Entry
Above: View north from entry

10
CHENG-BRIER HOUSE

10

成-布赖尔住宅
CHENG-BRIER HOUSE

TIBURON, CALIFORNIA
DESIGN / COMPLETION
2011 / 2016

The 6,000-square-foot Cheng-Brier House is located on a dramatic down-sloping hillside in Tiburon, California, with spectacular and unobstructed views of San Francisco, San Francisco Bay, and the Golden Gate Bridge to the south.

The home is organized on three levels that step down the steep hillside. The entry and all public spaces—kitchen, dining, living, and family—are located on the middle level and are accessed by dual landscape stairways from the east and west ends of the property. A double-height living space adds drama and vertical connectivity to the interior spatial composition. Family bedrooms are located on the upper level, with the master suite featuring a private outdoor Japanese *onsen*—open-air bathing area—that overlooks San Francisco Bay. A children's play area is located on the lower level.

Deep cantilevered terraces are located on the south side of all levels of the home, extending interior spaces to the exterior, and creating outdoor living areas with some of the most beautiful views in the San Francisco Bay Area.

On the middle level, a narrow interior-exterior water runnel slices through the home and connects a water feature at the entrance and the pool surrounded by cantilevered decks outside. Below the home, a series of pathways and rock-framed terraces lead to a pond at the bottom of the site.

The north side of the house has been designed with a strong rhythm of closely spaced teak vertical mullions, providing a sense of screened privacy from the uphill neighbors. In contrast, the south side of the home is clad in floor-to-ceiling glass, maximizing views of the San Francisco Bay Area.

The material palette is an elegant combination of bright white stucco, Windsor limestone set in a custom horizontal pattern, stained Western Red Cedar horizontal boards, and teak planks used for the rhythmic screen wall on the north-entrance side of the house.

Previous pages: House from below
Above: Living, dining, family and kitchen from pool deck

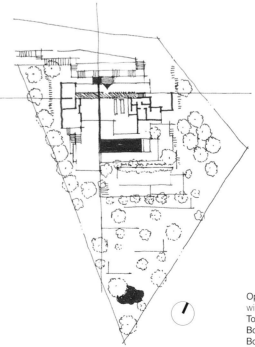

Opposite: Living room view of the bay, with San Francisco beyond
Top: Kitchen
Bottom left: Model
Bottom right: Site plan

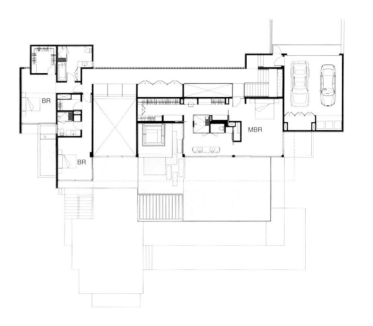
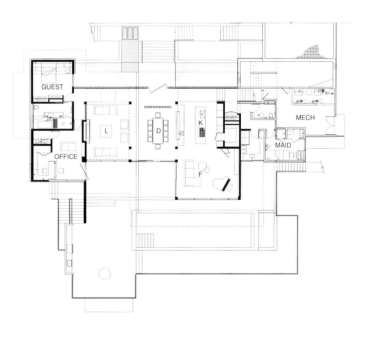

Opposite: Facing east from deck
Above: Floor Plans
Following pages: South elevation from below

123

11
KUENSTER-MIERS HOUSE

11

门斯特-迈尔斯住宅
KUENSTER-MIERS HOUSE

SONOMA, CALIFORNIA
DESIGN / COMPLETION
2012 / 2014

A familiar family story: the children grow up, leave for college, and pursue their own careers, leaving a wonderful but empty home behind. Architect George Miers and attorney Jenny Kuenster found themselves in this situation. While downsizing their spacious Orinda residence seemed obvious, their dilemma was how to maintain a sense of "home" their children and families would want to return to. They felt the nearby Sonoma/Napa Valley area offered a perfect solution: a home in the wine country that met not only their personal living criteria but held its own allure for their young adult children and accompanying entourage.

The initial program was to find a parcel of land large enough to construct a small home with a big kitchen for the entire family to gather for holidays, an ample area for entertainment, and a comfortable master bedroom: anything larger would be too much to clean every weekend. Additional living spaces were needed for the children, and consideration made for a caretaker, should one be required in the future. The program, therefore, evolved to include a pool with a pool house that could be converted to sleeping quarters with air beds, a small guest house, and an apartment that could be rented.

Miers and Kuenster found a magical property in west Sonoma, once the site of Napa/Sonoma's earliest vineyard. The 3-acre site featured a tired 1970s three-bedroom bungalow with a connecting garage, but it offered wonderful views, spectacular fruit trees, and an open meadow used by the ranch up the road to graze horses. The resulting design integrated program and budget. The existing 2,200-square-foot home was essentially gutted with the living/dining area rebuilt over the existing foundation to reduce costs. A 600-square-foot kitchen addition was made, and the overall exterior integrated into a modern indoor-outdoor living space featuring warm cedar siding accented against light cement-plaster walls and large planes of sliding glass doors. To further reduce costs, the bedroom wing was kept largely intact with two bedrooms combined into one comfortable master suite and a reconfigured light-filled master bath opening onto a private rock garden.

Completing the "compound," the old garage was converted into a guest house, a new three-car garage was connected to the main house via a covered cedar bridge, and a secluded one-bedroom apartment was added behind the garage. A pool and pool house was then sited to frame the main outdoor living areas while a 1.25-acre vineyard of syrah and roussanne was planted upon the former pasture. A vegetable garden is tucked into what had been a forgotten corner of the property with six large, raised planters surrounding an outdoor dining table.

Previous pages: House sitting beyond the vineyard of syrah and roussanne
Above: Pool and pool house
Following pages: East elevation of main house

Above: Site plan
Right: Living room and entry

12
CHENG-REINGANUM HOUSE

12

成-雷因格纳姆住宅
CHENG-REINGANUM HOUSE

ORINDA, CALIFORNIA
DESIGN / COMPLETION
2013 / 2016

The Cheng-Reinganum House is located at the top of a beautiful hillside property in Orinda, California, surrounded by mature oak, pine, and sycamore trees. Views of the rolling hills and Mount Diablo beyond are to the south and east. The owners, a family with two children, lived nearby for many years, and with the children close to college age, they embarked on creating a new home to reflect their new lifestyle: single-level open planning, seamless connections to the outdoors, space for the children to call "home" when they return from college, and a lower-level guest suite for visiting parents.

The existing Mid-Century home on the site had an interesting plan organization, but it suffered from very low ceilings and eaves (6 feet 4 inches) that effectively blocked almost all views and created dark and uninviting interior spaces. The design strategy involved creating the home over the footprint of the pre-existing building and using 90 percent of the original foundations for economy. Higher ceilings and plenty of daylight were a must. Family living is on the ground level, while the garage and a small guest suite are located on the level below.

The plan is based on three wings set at 105, 120, and 135 degrees apart from each other. The east wing includes the living and dining spaces; the south wing includes the kitchen and family room; and the west wing includes the master suite. All major spaces, including bedrooms, have direct access to the outdoors. Floor-to-ceiling glazing provides strong indoor-outdoor connections, and carefully placed clerestory windows serve the home with dappled light from above, while affording views of sky and treetops.

The landscape has been designed for informal family living, with a large, linear pool sited between the family and master-bedroom wings. The living and dining wing opens directly to a large deck that dramatically cantilevers into the canopy of a beautiful California live oak. Watching the sun set over the hills from this deck is a dream come true for the homeowners.

Previous pages: Entry elevation
Above: Entry approach

Opposite: Living room, facing east
Left: Light filtering into hallway
Right: Master bedroom looking out towards pool deck

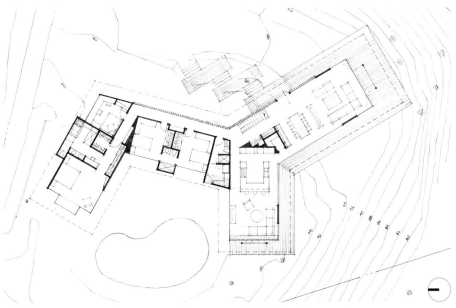

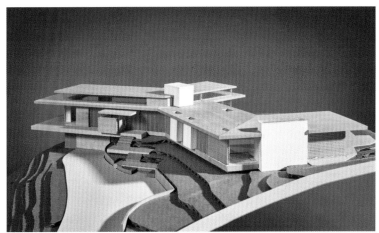

Opposite: View of kitchen and family room opening up to pool
Top: Approaching entry at dusk
Bottom left: Site plan
Bottom right: Model bird's eye view

13
MORA ESTATES

13

莫拉地产
MORA ESTATES

LOS ALTOS HILLS, CALIFORNIA
DESIGN / COMPLETION
2011 / 2016

Located at the end of a cul-de-sac next to an open space preserve, the Mora Estates are a special enclave of four high-end Silicon Valley houses for sale. Lot 2 is the first completed of the modernist cluster; the developer plans to complete three more homes with similar architectural language and materials.

In the heart of Silicon Valley, this house boasts an amazing view of the Bay, and the rolling hills. The home is organized to best frame the view to the Bay in three parallel bands. Past a small private vineyard, the gracious entry court ushers you in under a low protective eave. Beyond, the dramatic double-height space opens up to a dining room and then steps down to the living room and the pool beyond. To the left is the kitchen and family room, both with access to the view, and to the right is the first of two master suites, and a more intimate library with its own patio. The palette of dark brown woods with marble floors and white walls gives a distinctly glamorous feel to this home.

Upstairs a second master suite has its own balcony and three children's bedrooms are on the other side of the open space. Downstairs a day-lit recreation room with a wine cellar along with a home theater and gym offers every amenity.

From the pool / spa terrace, a pathway leads past the infinity edge of the pool to the guest house, beyond which terraced lawns step down through the oaks.

Previous pages: East elevation
Above: Library and south courtyard

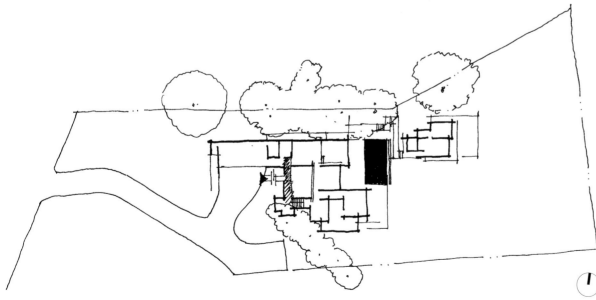

Opposite top: Living and dining room, facing entry
Opposite bottom: Master bathroom
Top: Entry approach
Bottom: Site plan

在建住宅

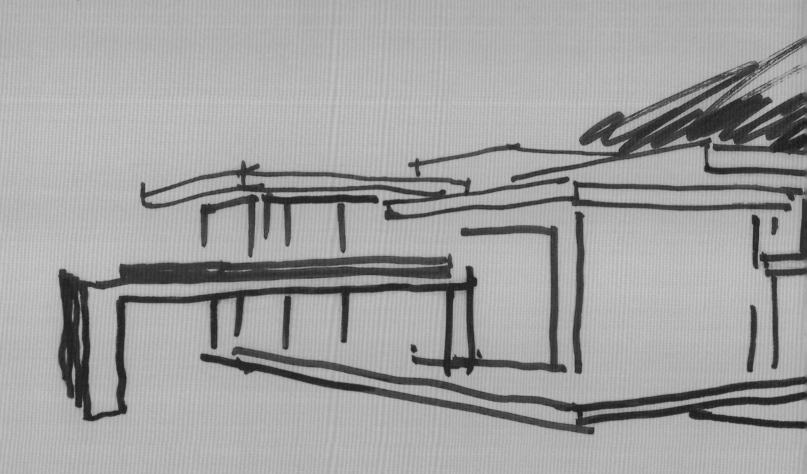

HOUSES
IN PROGRESS

14

印度北部住宅
HOUSE IN NORTHERN INDIA

NORTHERN INDIA
DESIGN 2012

In 2012 Swatt | Miers was commissioned to design a new grand estate in Northern India. Three members of the office traveled to India to gain an understanding of the owner's goals, the context of the site, and local building methods, and spent three weeks conceptualizing the design in a conference room that had been converted into a temporary architecture office dedicated to this project. The architectural team met with two Vaastu masters to better understand Indian Vaastu traditions, as well as with the landscape architect, and mechanical and structural engineers. They prepared conceptual plans for the new home and built a scale model of the design. Over the following six months, the design was refined and completed in the Swatt | Miers office in California.

The site is a flat, 5-acre walled "farmhouse" parcel located near a transit hub and a major city. Other than rows of neem trees adjacent to the perimeter walls, the land is devoid of vegetation. Vaastu traditions shaped the grading of the land to a high side on the southwest and open space to the northeast.

The design that evolved is a 27,000-square-foot, two-story rectangular structure that reaches a height of 40 feet. The plan is organized around a central double-height skylit Brahmasthan, which according to Vaastu, the ancient tradition of architectural and planning principles, is the holiest and most powerful zone of the residence. Due to the heat and dusty climate, the typically open-air Brahmasthan was transformed into a garden atrium with a sawtooth roof that faces north so sunlight enters during the winter months and is excluded in the summer. The kitchen, dining, master bedroom, and study are all in carefully prescribed locations around the Brahmasthan as mandated by the Vaastu masters. A swimming pool and detached pool house are located to the north and northwest respectively. A bridge from the second floor connects to the southwest berm and provides access to the gardens.

In keeping with Indian traditions, a low stone wall separates the main parcel from an adjoining secondary parcel to the south. Accessed by a low bridge that crosses the wall, the secondary parcel has been developed for entertaining, with large expanses of lawn and a unique one-story entertainment pavilion that appears to float on a large body of water, bordered by an allée of trees.

Above: Entrance elevation

Top: Concept sketch
Bottom: Model view of entrance
Opposite: A low bridge provides access to the entertainment pavilion

15 圣地亚哥住宅
CASA SANTIAGO

ATHERTON, CALIFORNIA
DESIGN 2014

Casa Santiago is located on a prominent corner parcel in Atherton, in the heart of Northern California's Silicon Valley. Accessed from a quiet street on the west side of the property, the project has been designed to utilize every square inch of its 1.5-acre site. This new 13,500-square-foot home is organized around a U-shaped plan. It separates the private and public spaces within the house and celebrates connections to the landscape with large areas of operable glazing and access to multiple "outdoor rooms" and other functional areas within the site.

Outdoor areas include a Zen garden between the wings of the house, an exercise terrace adjacent to an indoor gym on the north side of the property, a sculpture garden viewed from the dining room, a sunken garden with water and fire features, and a tennis court. A dramatic 70-foot-long colonnade, shaded by a linear trellis, connects the main house to an accessory structure located at the southeast corner of the property. It contains a four-car garage and a skylit 35-foot-long indoor swimming pool.

Above: Courtyard view

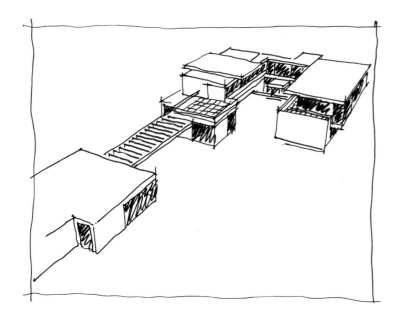

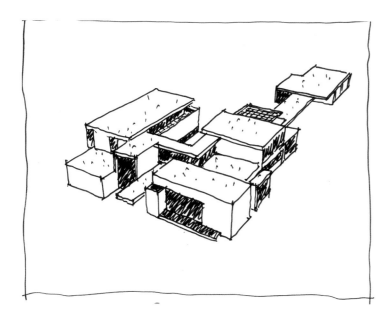

Opposite: Concept sketches
Top: View of entrance
Bottom: Model bird's-eye view

16 太阳神之屋
HELIOS HOUSE

CALISTOGA, CALIFORNIA
DESIGN / COMPLETION
2013 / 2017

Designed for indoor-outdoor living and maximizing views of vineyards in the foreground and mountains in the distance, this new 6,300-square-foot home is envisioned as a multi-generational gathering space for a young couple with two children and family members in the surrounding area. An existing home on the site, very visible from the busy Silverado Trail, was poorly located and in a low zone where it missed out on incredible views available from the higher portions of the 7.5-acre site.

The new home is located on top of a knoll previously occupied by a swimming pool. The new site shields the home from the roadway, while focusing views to vineyards and rolling hills to the west. Accessed from a new driveway and entrance courtyard on the north side, the home's circulation spine runs east to west, aligned with the top of the ridge. In counterpoint, the master suite on the north side and the guesthouse on the south side run perpendicular to the grade, dramatically cantilevering over stone plinths into the treetops of existing madrone, pine, and heritage oak trees. Built of stone, wood, and glass, this new home is a careful modern insertion into a sensitive wine-country environment, preserving all but a few of the hundreds of existing trees on the property.

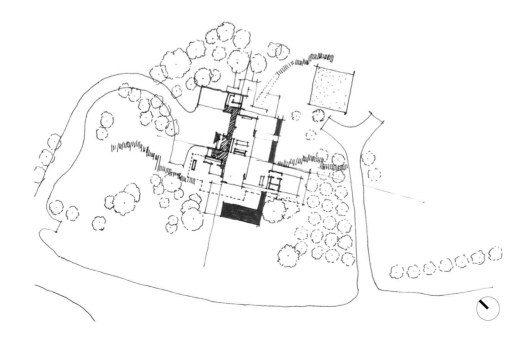

Opposite: Site plan
Above: View of cantilevered guest house from below

159

Left (top, middle and bottom): Model views
Above: View of south terrace and guest house

17

阿玛拉住宅
AMARA HOUSE

ATHERTON, CALIFORNIA
DESIGN / COMPLETION
2014 / 2017

This new family home in Atherton is located on a flat corner lot with beautiful mature trees along the perimeter of the property, affording privacy on all sides.

The L-shaped plan of the main house has public spaces on the first floor opening up to expansive landscaped areas and a swimming pool. Deep cantilevers with trellises provide sun protection for the interiors and create beautiful dappled light patterns for the outdoor spaces. Exterior forms are simple compositions of stained Western Red Cedar horizontal boards and integral colored stucco. The second floor at the ends of the 'L' cantilever over the first floor, while a vertical stair tower faces the main street, Atherton Avenue, and is a visual counterpoint to the horizontality of the overall design.

The project also includes a detached garage that interlocks with the main house, and a detached guest house that continues the strong line of trellises from the main house and anchors the short end of the linear pool.

Above: North elevation

Opposite: Floor plans
Top, middle and bottom: Model views

18 私语之石住宅
WHISPERING STONES HOUSE

HEALDSBURG, CALIFORNIA
DESIGN / COMPLETION
2014 / 2017

The design of Whispering Stones House is about the harmony and balance of architecture and nature. An untouched and heavily wooded 38-acre site in the hills above the Northern California wine-country town of Healdsburg is the setting for this ultra-sensitive architectural insertion into nature.

The owners, a couple moving to the San Francisco Bay Area from Montana, desired a modern home that combined rigorous architectural design with a reverence for the natural setting. The design is a linear, three-level bar that runs north to south over a knoll at the top of the site, perpendicular to the contours of the land. Anchored in the center at the top of the knoll by a cast-in-place concrete stair tower, the home dramatically cantilevers off of the knoll to the north and south and into the treetops of the untouched forest. Terraces accessed by the main living spaces are perpendicular to the main axis of the house, and a linear swimming pool follows the ridge of the land to the east, focusing views on rolling, oak studded hills in the distance.

Built primarily of cast-in-place concrete, wood siding, and glass, the architecture is both assertive and, at the same time, in perfect harmony with the beautiful California wine-country environment.

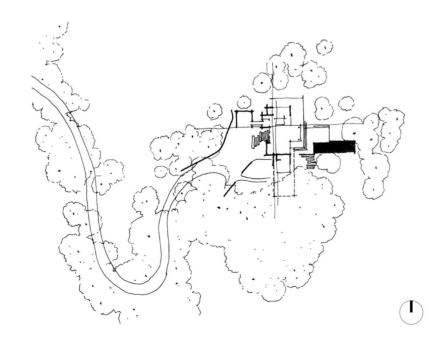

Opposite: Site diagram
Above: East elevation

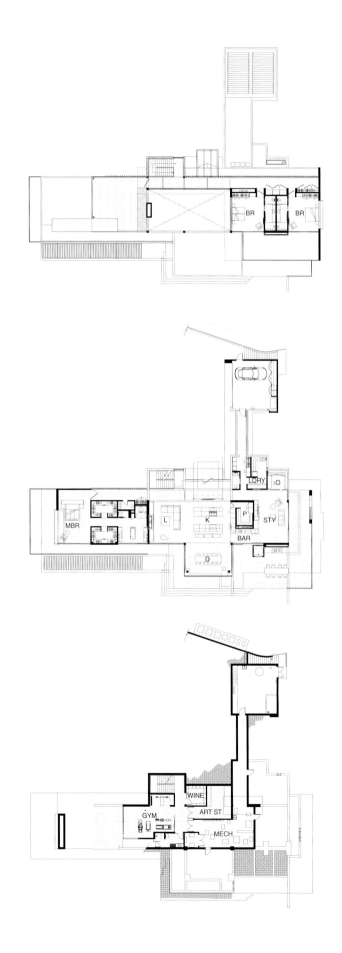

Opposite left: Floor plans
Opposite right: Model view
Above: View of entrance from the west

19 入江住宅
IRIE HOUSE

BALEARIC ISLANDS, SPAIN
DESIGN / COMPLETION
2014 / 2018

In the spring of 2014 Swatt | Miers was asked to design a new home and two levels of an adjacent four-story apartment building on the top of a steep cliff in the Balearic Islands, overlooking the Mediterranean Sea. The site is narrow and relatively flat, bordered on the west side by the four-story apartment building, and on the east side by a heavily wooded agricultural preserve. The southern edge of the site is the top of the cliff, 22 meters wide and overlooking the ocean. A steep stone stairway carved into the side of the hill, leads to a private beach 35 meters below.

Because of the linearity of the site, the garage structure and house have been separated, with parking located adjacent to the street on the north side of the property. A linear bridge runs parallel to the landscaped entry courtyard, terminating at the entrance to the home. The main house structure has been placed as far to the south as allowed by code to ensure the master suite on the second floor and the kitchen, dining, and living spaces on the first floor enjoy magnificent, unobstructed views of the ocean and beautiful nearby island. Connections to the outdoors are seamless, with a large stone terrace and swimming pool located on the south side of the house, perched above the ocean at the edge of the cliff.

Opposite: Model view
Above: View of the house from the north, overlooking the Mediterranean Sea

Left: Model view of house on cliff
Above: Pool, terrace, and south side of house

20

欧西欧里住宅
ORCIUOLI HOUSE

ATHERTON, CALIFORNIA
DESIGN 2013

Located on a relatively flat lot in Atherton under a sheltering canopy of mature oaks, this single-story family house with basement, forms a U-shape around a strong central axis. The homeowners came to Swatt | Miers with an appreciation for the Case Study Houses of Southern California. They loved the clean lines and the clear logic of their plans. A stone stair tower defines the vertical, while long linear white roofs enclose the three wings of the home. A trellis shoots through the entry door and main living space, out into the courtyard where a pool continues the axis and ends at a water feature emerging from the pool house wall.

The main entertaining spaces are parallel to the street. One wing houses family spaces—the kitchen and playroom—and the wing on the opposite side of the pool houses the bedroom areas. The pool house, guest house, and shady barbecue terrace are formed by a linear detached structure in counterpoint to the main axis.

Above: Front elevation

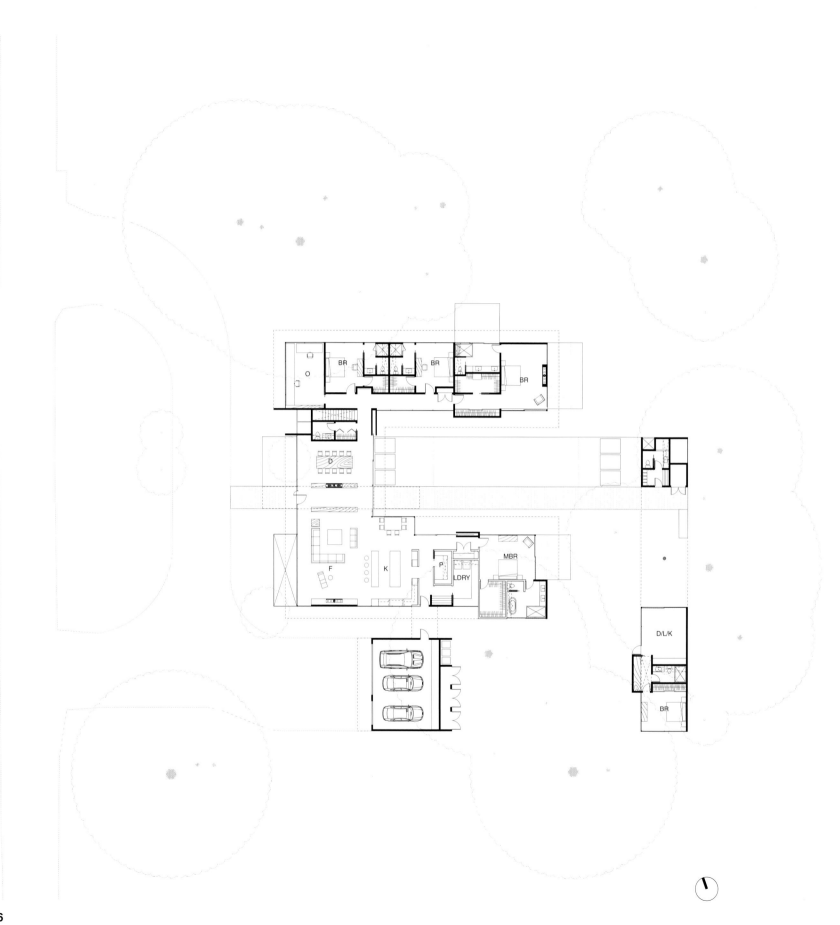

Opposite: Site and floor plans
Top: Courtyard view
Bottom: Model view

21 西风住宅
WESTWIND HOUSE

SILICON VALLEY, CALIFORNIA
DESIGN 2015

Westwind House is located in Northern California's Silicon Valley, on a beautiful hillside property with views of the valley below and San Francisco Bay in the distance. The design is based on an H-shaped plan with a double-height living/dining pavilion located at the center of the composition, flanked by a two-story family wing on the north side and a one-story office/recreation/gym wing on the south side. A floating bridge traverses the great room and connects the family wing to a rooftop terrace that sits above the west wing and provides a tranquil place to enjoy views of the surrounding hillside. The orientation of the home has been designed so the great room, master suite, kitchen, and family room directly face San Francisco Bay.

Glazing has been designed to provide a sense of privacy at the entry courtyard, with floor-to-ceiling wood-screen windows that minimize views to the interior. In contrast, the east side of the house has been designed almost entirely with walls of sliding glass doors that open the home to the beautiful views and the landscape, including stone terraces, outdoor kitchen, swimming pool, spa, and expansive play ares for the children.

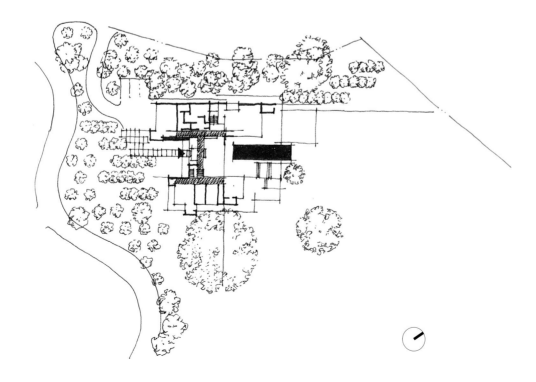

Opposite: Site plan
Above: Entry courtyard

Left: View from northeast
Top: Model view from southwest
Middle: Model bird's-eye view
Bottom: Model view from east

22

查隆路住宅
CHALON ROAD HOUSE

LOS ANGELES, CALIFORNIA
DESIGN 2014

Swatt | Miers first new residential project in Los Angeles commenced construction in June 2016. Located in the prestigious Bel Air Hills, the site of this two-story 9,000-square-foot home presented a number of difficult challenges: a relatively small and constrained buildable area from which the remainder of the site rises up at a 45-degree angle; geotechnical investigations that revealed traces of a fault zone running through portions of the site requiring a 160-foot long by 12-foot wide soil nail retaining wall to resist forces from the steep hill behind; and the City of Los Angeles's set back and height requirements that combined with the resultant small site area defined a strict three-dimensional envelope into which the home had to be sculpted. The design, therefore, with its stepped levels and sweeping second-floor curved form is a direct response to the city's zoning envelope.

Despite the limited building area, the home is sited so as to establish three distinct outdoor areas—vehicular-entry courtyard, central garden court, and a second-floor bedroom garden that engages the dramatic hill behind. Upon entering the site, one's first experience is the elegantly landscaped vehicular-entry court featuring a two-car garage and four tree-shaded visitor spaces. The ambience of this carefully detailed space is defined by the integration of special paving, decomposed granite, and drought-tolerant landscape further accented by the soothing sounds emanating from a linear water feature that welcomes visitors as they cross over to enter the home.

The first floor of this six-bedroom home features a two-story living/dining area that opens onto the garden courtyard and designed for year round indoor-outdoor living. A breakfast room/kitchen/family area also opens onto the garden, while a guest suite adjacent to the garden contains its own private patio. The first floor also features a home theater, wine cellar, and parking garage.

A striking open-tread stair with glass railing leads to the second floor. While this floor overlooks both the garden and vehicular courtyards, each bedroom has its own private garden away from the public areas. In addition, this floor features an office, indoor-outdoor sitting areas, and a two-story gym with a spiral staircase that accesses a dramatic roof terrace.

The stepped exterior design integrates light-colored stucco walls with stone and cedar accents that together, present a warm modern expression unique to the area.

Above: East elevation

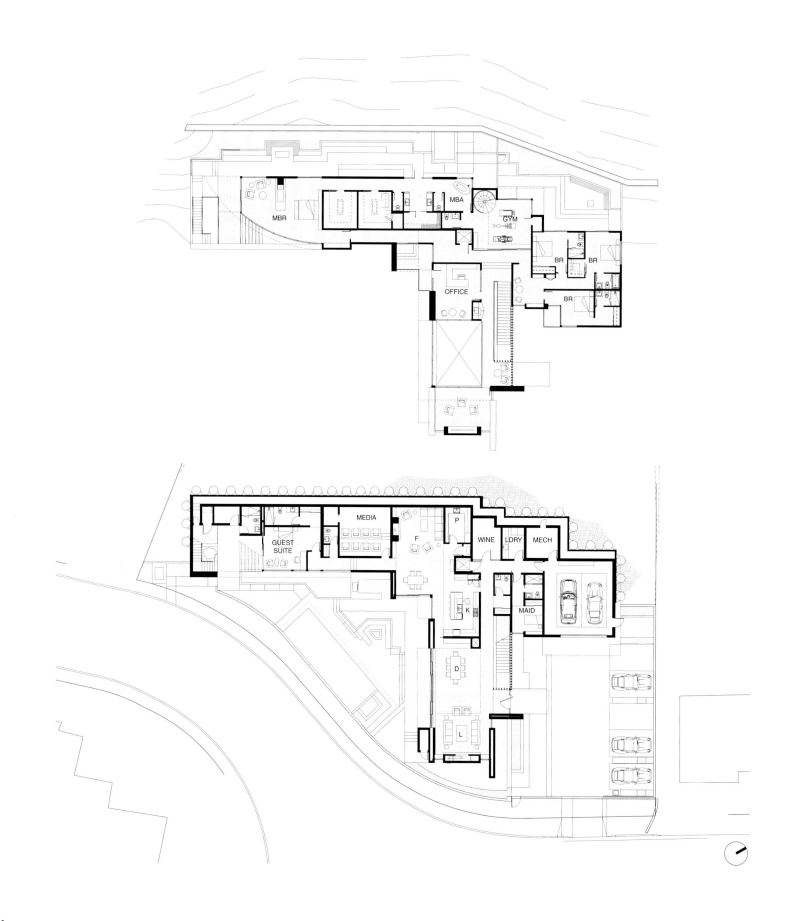

Opposite: Floor plans
Above: View of entry approach

23 李住宅
LEE HOUSE

LAFAYETTE, CALIFORNIA
DESIGN / COMPLETION
2015 / 2018

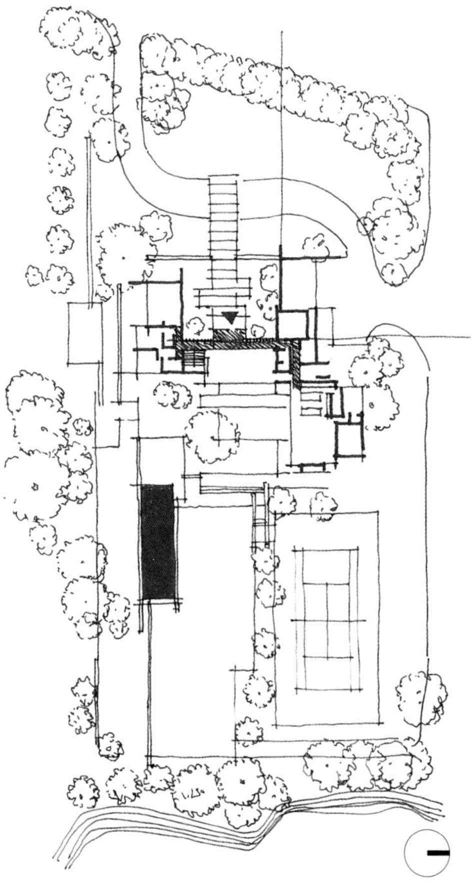

Lee House is located on a 2.3-acre estate parcel in the Happy Valley neighborhood of Lafayette, California. The relatively flat site is bordered on all sides by mature landscaping, including California live oak and redwood trees, tall English Laurel, Privet and Oleander hedges, and a year-round creek. This new home replaces a 1938 home and is substantially within the previous home's footprint, thereby protecting and maintaining much of the preexisting landscape.

The new 10,000-square-foot home is organized in a H-shaped plan with a large double-height great room in the center of the composition, flanked by two-story wings. On the first floor, the kitchen and family areas are located in the east wing, while a private office, guest suite, and gym occupy the west wing. Upstairs, the children's bedrooms are located above the family wing, while the master suite is designed as a private retreat, with a separate office in the west wing. A floating bridge pierces the great room and connects the two wings on the second floor.

Exterior landscape features include a new tennis court, various lawn and garden environments, and a new swimming pool oriented north-south. The swimming pool and adjacent terrace are partially covered by an asymmetrical shade structure supported by two offset linear walls.

Opposite: Site plan
Above: South elevation

187

Left: View out from great room
Top: Model view of entry court
Middle: Model bird's-eye view
Bottom: Model view of south courtyard

24 河畔住宅
RIVER HOUSE

FRESNO, CALIFORNIA
DESIGN / COMPLETION
2015 / 2018

The owners, a young couple with two children, contemplated building a special family home for years. The site they chose is a 20-acre flat agricultural parcel of almond orchards that overlooks the San Joaquin River in Fresno, California. It is accessed by a new private road that winds through the almond trees, arriving at an entrance courtyard on the north edge of the orchard, adjacent to the river

The plan is a simple T, with a south-north wing containing the garage and all of the service spaces, and a two-story east-west bar that includes all of the public living spaces on the first floor and all of the family private spaces on the second floor. A linear circulation spine runs along the south edge of the bar so that all living spaces enjoy beautiful views of the river. A swimming pool and pool house structure are located to the east of the main house, further stretching the linearity of the architectural composition. The design includes floor-to-ceiling glazing in most living spaces, deep roof overhangs to provide shade protection from the extreme Central Valley heat, and generous new landscaping that ties together the agricultural qualities of the site with the residential character of this riverfront home.

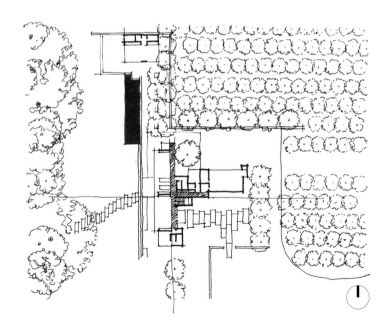

Opposite: Site plan
Above: View from river's edge

Left: East elevation
Top: Model view from northwest
Middle: Model view from southwest
Bottom: Model view from river

25 黑点住宅
BLACK POINT HOUSE

HONOLULU, HAWAII
DESIGN 2015

For 30 years, the owners resided in a vernacular Hawaiian home on the site, a windy hilltop overlooking Koko Head. Their existing home featured a very tall gabled roof with low eaves, and unusable balconies due to high winds. Wishing for a bright and open modern home, they invited Swatt | Miers to redesign their house from the ground up, with light interiors, great views, and better protection from the wind.

Sheltering stone walls embrace the home, protecting it from the strong northerly trade winds that gust through the site. Two parallel stone walls shoot toward the view and full-height glass walls enclose the great room and master bedroom providing beautiful views of Koko Head and Maunalua Bay. These walls are anchored on the property by a stone tower that sits on the high side of the site, and connects to a tall circulation spine. High roofs with wood ceilings float above the stone walls.

Opposite: Site plan
Above: Entrance and lanai overlook the ocean

195

Opposite top: Southeast elevation
Opposite bottom: Living room with view of ocean
Top, middle and bottom: Model views

26 凯路亚海滩别墅
KAILUA BEACH HOUSES

O'AHU, HAWAII
DESIGN 2014

This dramatic beachfront lot on O'ahu is the last place Elvis Presley vacationed in Hawaii. The client asked Swatt | Miers to develop schemes for multiple houses—all with views—to feel like separate properties with gardens and pools. The lot is relatively flat except for a small berm on the beachside of the property. In this scheme, one slightly larger home gets the premier position at the front of the lot, while all three houses are angled to take advantage of the ocean view.

All of the homes feature a raised floor that hovers over the site for an improved view. Exposed-concrete towers enclose stairs and provide support on either side of the homes, holding the floating slabs of floors and roof. Dramatic glass walls direct the gaze to the view, while the sides and fronts are more opaque for privacy.

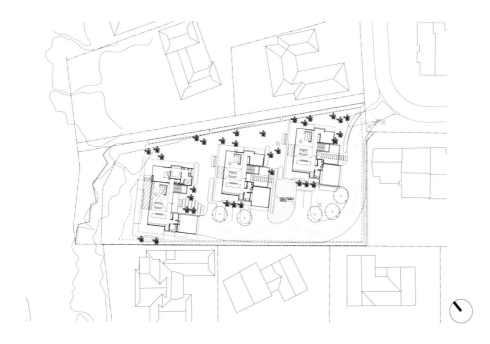

Opposite: Site plan
Above: View of three houses from the beach

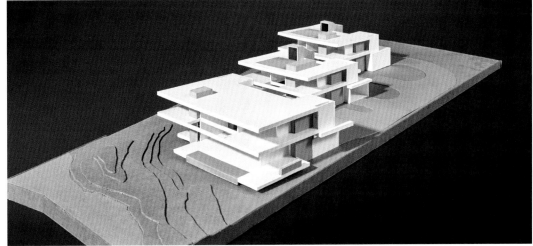

Left: Entry of front beachside house
Top: Bird's-eye view
Bottom: Model bird's-eye view

27 萨那住宅
SANA RESIDENCE

LOS ALTOS HILLS, CALIFORNIA
DESIGN 2016

SANA Residence is sited on a steep northwest-facing hillside parcel in Silicon Valley's semi-rural community of Los Altos Hills. It is accessed by a panhandle driveway from the northeast.

In response to the town's strict limits on "maximum development area," the design solution is a compact, linear three-level bar that parallels the contours of the land. All major spaces are oriented for downhill views, with large areas of glazing on the uphill side of the home designed to maximize natural lighting from the south. A double-height, wood-clad family/great room is the central space in the home and is surrounded by white cement plaster-clad volumes on the ends of the long bar design. Rectangular cement-plaster wall, floor, and roof extensions visually connect interior spaces to the exterior and frame views of the beautiful landscape and valley beyond.

Above: Concept sketch

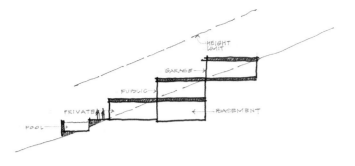

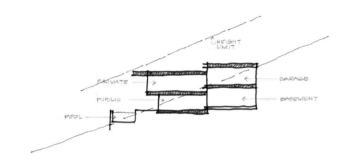

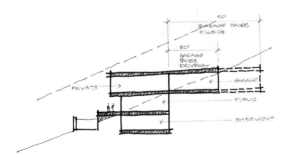

Left: Concept sectional diagrams
Above: Northwest elevation

商业、教育与机构项目

COMMERCIAL, EDUCATIONAL & INSTITUTIONAL PROJECTS

从地点与客户需求中产生的建筑

AN ARCHITECTURE EMERGING FROM SITE & CLIENT NEEDS

George Miers, AIA

I have always loved solving problems.

As a child, crossword and picture puzzles were constant companions. So in hindsight, it seems natural that I was attracted to building types with complex internal program needs such as libraries, police facilities, hospitals, and animal shelters. These are buildings where the operational requirements and spatial adjacencies are critical for the building to function as desired, and the designer needs to assemble all the disparate parts and meld them into one unified whole. I find this to be the ultimate problem-solving puzzle; a challenge wherein one first defines all the pieces and then must assemble them in a manner that ultimately supports, and actually dictates, operational activities within. Of course the art of this puzzle solving is when the end product blossoms into a magnificent structure that has evolved from inside out, expressing its complex internal organization. For me, this is the joy and reward of architecture.

Architecture is a journey

To experience this joy and to reap the rewards, one must first develop an inherent understanding of and affinity for the subject matter—architecture. Growing up in San Francisco's Visitacion Valley, a working-class neighborhood sandwiched between Cow Palace and Candlestick Park, did not afford many good examples of modern architectural design. While I had heard the name Frank Lloyd Wright from my Great-Uncle Charlie, a carpenter who constructed homes in Half Moon Bay, I had no frame of reference for Wright's work, nor anyone else's. Ironically, the only significant project constructed in my neighborhood during the 1950s and 1960s was Joseph Eichler's first and last low-income high-rise project, Geneva Towers, which rapidly deteriorated into a social catastrophe, not unlike the problems that faced other low income projects during the 1960s. The net result was that, for years to come, the name Eichler did not conjure up the same images of wonderful, modern, open-plan living environments for me as it did for other budding architects of my generation, such as my incredibly gifted partner Bob Swatt. Thus, upon departing for college in 1967, I had not only a lot to learn, but a lot of preconceived images to shed. And the latter took some doing.

Attending architecture school during the "power to the people" generation in the late late-1960s and early-1970s—at Washington University in St. Louis and later University of California, Berkeley—certainly fueled my interest in user needs as the underlying basis of good and responsive architectural design. It also spurred my interest in affordable housing, which was the predominant design assignment during the 1960s and early-1970s. It was a building type to which I could relate, and for which I found my problem-solving skills were right at home developing well-thought-out, minimalist, low-income designs and assimilating complicated program requirements into socially significant building types. Most of these design assignments came easily, but towards the end of my tenure at Washington University I hit an unexpected bump in the road on an amazing summer program in San Miguel de Allende. An incredible faculty, led by Ricardo Legorreta, Charles Moore, and the brilliant Swiss-architect Dolf Schnebli, presented a relatively simple design problem: design a second-home "resort" community in the surrounding San Miguel hillsides and do so with simple, indigenous materials. There were no specific spatial requirements; in fact the idea was that land and labor were relatively inexpensive so one could afford to splurge with larger, spacious units. In short, we were to make up our own program that would define "good living." While my classmates loved the project—at least those who showed up and were not hanging out at La Cucaracha, the local bar back in town—I hit a wall. No matter how hard I tried, I could not get away from minimalizing every living space, which of course was not the assignment. It was a moment to reflect upon as I had no frame of reference to design an environment like this, having never lived in or even visited such a community. The lesson learned was that clever problem solving wasn't enough. Architects need to understand the environments they are designing and have an affinity for the clients who will occupy them. I left San Miguel committed to changing that.

Following this experience, I returned to San Francisco and was fortunate to land at the dynamic office of Skidmore, Owings & Merrill (SOM) under the leadership of Charles Bassett. This was my first serious design firm experience; and while incredibly stimulating to be amongst such an outstanding collection of designers, it was also a bit nerve racking working, again, on building types I had limited experience with and materials I had little knowledge of. The SOM building rules were fairly simple to learn but when I was placed on a project for the design of a new hotel on the Island of Hawaii (Big Island)—a follow up to SOM's award-winning Mauna Kea—I had to ask for time off so I could visit the original. Thus commenced a period of significant travel visiting and experiencing as many important buildings as possible.

My years at SOM instilled an appreciation for the depth of detailing needed to execute a truly elegant building. It also provided a basic understanding of how buildings were put together. However, it was my next two positions that allowed me to start assembling all the pieces together and further honing my problem-solving skills. The first was a position with Sim Van der Ryn, the new State Architect, in which I developed the facility program for the California's first Energy Efficient Office Building along with many other state buildings. Here I learned the art of programming from the very best—Bobbie Sue Hood; how well-documented user requirements can plant the seeds that should blossom into an efficient and responsive architectural plan was a profound lesson. Hood also instilled a sense of pragmatism; "make sure all the proposed criteria (adjacencies, building areas, etc.) actually work before documenting them" and "never list options if you don't agree with them." In other words, the architect needs to test all user requirements as the program is being developed to make sure they really work and to have a vision of the final end product—a rule I follow to this day. My tenure at the State Architect office was followed by a four-year stint as a Commercial Studio Director at Kaplan McLaughlin Diaz Architects (KMD) where I was finally able to meld all the pieces together from program development to design and finally construction. Herb McLaughlin, in particular, took me under his wing and gave me a lot of freedom at a very young age, for which I will always remain grateful.

When I started my own practice in San Francisco in 1982, the majority of my work was focused on small speculative office buildings on relatively flat suburban sites. If my years at SOM and KMD taught me anything, it was the formulae for lease depths and elevator/stair/restroom central support cores, which have their own intrinsic logic. For a while I found myself intrigued with the challenge of solving the most efficient office-building puzzle. I reached the pinnacle of this effort in the early-1980s when pushed by my client to develop an office design layout that was 102-percent net leasable based upon a concept of designing the building's circulation as covered exterior walkways which, while not factored into the building's gross square feet, fell within the definition of leasable. My client loved the clever solution but there was a limit to the satisfaction I was gaining from "minimal program" commissions like this. As I looked around at building types that seemed more appealing, I found many of them were in the public sector—libraries, city halls, police facilities. Knowing the likelihood of obtaining such a commission with little experience was slight, I entered a competition for the Novato Civic Center, which I won. While this project ended up stalled due to political reasons, it led to the award of my first constructed public building, the Dublin Civic Center, which included a city hall, council chambers, and police facility. The five council members not only proved to be great clients but long-lasting friends whose recommendations assisted me to land numerous other high-profile public facilities, including the Livermore library, police, and civic center facilities; the Santa Cruz, Antioch, and Lodi police facilities; the San Diego Campus for Animal Care; and the San Carlos Library. The intricate internal programs in these projects provided me with the problem-solving challenge I had longed for.

Hooking up with Bob
Over the years, my practice moved from San Francisco to Oakland and eventually Moraga, a few blocks from my home. As the children moved on to college, there was not much reason to be located so far away from the hub of architectural practices. I was feeling the daily need to interact more strongly with colleagues, which led me to find a partner. I had always admired the work of Robert Swatt—his crisp modern designs and, in particular, his uncanny ability to knit buildings to very difficult steep sites such that they seemed inevitably there. However, as we discussed mutual interests in bringing our practices together, I found an even stronger bond. Not only did we share an unwavering commitment to designing buildings from the inside out, but I found Bob did so with incredible respect and deference for his client's programs, welcoming their input and ideas, and believing the resulting, more complex program made for a stronger design. Our common interest in this design and program approach has made the past eight years the most rewarding of my career; and as we look to the future, we feel the potential is limitless.

Moving forward with our clients

An architect's job is more than creating beautiful, exciting edifices that express a personal aesthetic bent. In the end, we are commissioned to evolve a design that satisfies our client's program needs and solves their specific building problems. The first step in that process is to correctly define the actual problem requiring a solution. While that may sound obvious, all too often building design solutions are developed from incorrect program criteria because the actual problem is defined incorrectly, resulting in an eventual building that might look good but doesn't function well. The myriad of award-winning public-housing projects of the 1950s and 1960s that had to be demolished in the 1970s is a prominent architectural example.

- Faulty information that leading to unfortunate decisions is a universal problem across disciplines, but in architecture it tends to happen generally for one of three reasons:
- the architect is too preoccupied designing from the outside in and misses or ignores criteria that should be molding the design from the inside out
- the architect convinces the client as to the wisdom of certain design solutions and the client either doesn't realize the everyday/operational implications or is too embarrassed/intimidated to speak up
- the architect doesn't probe the issues deeply enough with the client and at key junctures doesn't question the validity of the program criteria provided.

This last issue is a particularly slippery slope in that it requires respectfully balancing client input with the architect's own experience and common sense. I have often watched clients nodding in agreement to ideas I have suggested when my gut instinct tells me they really don't understand or agree with them. Similarly, I often find clients—from home owners to librarians to police officers—describe their desired criteria for a new building based on their current way of doing things because that is their limited perspective of the possibilities. An architect has a responsibility to their client to speak up when they sense a given direction may be in error; but doing so in a respectful manner is an art. While many architects dislike such interactions, we find them to be essential—typically when you have the key decision makers in the room rendering decisions that significantly influence the future design. This situation is when an effective architect is needed most—one with good common sense, peripheral vision, and strong design instincts who can sense the nuances of what the client is saying and constructively guide the program-definition process in a way that leads to the right design criteria. These client dialogues, which require careful listening and documentation—activities that many designers see as "non-creative" and "boring"—are the moments from which great user-responsive architecture is born.

An inevitable byproduct of architecture is that, good or bad, buildings tell a story about the client, architect, and era in which they are designed and constructed. While architects differ on the focus of this story, Swatt | Miers is committed to celebrating the integration of our client's program with the unique characteristics of the site and doing so in a manner that produces beautiful modern buildings with a timeless spirit.

28
PALM SPRINGS ANIMAL SHELTER

28

棕榈泉动物收容所
PALM SPRINGS ANIMAL SHELTER

PALM SPRINGS, CALIFORNIA
DESIGN / COMPLETION
2009 / 2011

Palm Springs Animal Shelter celebrates Swatt | Miers's affinity with domestic animals while integrating the city's long-held love affairs with Mid-Century Modernism and eye-catching modern art. (Palm Springs is, of course, the setting for Richard Neutra's iconic Kaufmann House and Liberace's piano-shaped pool.) Featuring a simple yet powerful cantilevered-roof overhang punctuated by palms, the public adoption gallery's colorful interior reaches out into the desert landscape and embraces the rugged San Jacinto Mountains through a large expanse of shaded glass. Set within the modern window wall that frames the indoor-outdoor space is a Mondrian-styled box punctuated with small windows and Corian® pedestals, which houses the feline adoption center. This play between program, environment art and architecture underlies the design of this desert jewel.

Located on a 3-acre site across from the city's popular Demuth Park, the 21,000-square-foot project represents a unique funding and operational partnership between the City of Palm Springs and the nonprofit Friends of the Shelter. Programmatically, the project's design features a centralized indoor-outdoor kennel arrangement with public adoption access provided within a shaded open courtyard/garden setting. This state-of-the-art facility includes secure animal control work areas, separate public intake and adoption lobbies, a training room for educational and evening uses, a large canine socialization room, and a fully equipped clinic for in-house medical pprocedures—all housed within a design that reflects the regional character of the unique Palm Springs community. The project has been designed as an equivalent LEED Silver facility and the wonderfully whimsical cat and poodle sculptures are by local artists Karen and Tony Barone.

Previous pages: View from street
Above: Entry approach

Opposite: Site plan
Top: Lobby space
Bottom: Canine area

Opposite: Public entrance
Top: Exterior of adoption courtyard
Bottom left: Mondrian-styled wall of cat adoption center
Bottom right: Façade detail

29
BROADWAY TENNIS CENTER

29

百老汇网球中心
BROADWAY TENNIS CENTER

BURLINGAME, CALIFORNIA
DESIGN / COMPLETION
2012 / 2015

Situated on a 2.8-acre former rental-car lot alongside Highway 101, this unique indoor tennis facility's impressive fabric-clad structure lights the night sky like a beacon. The 54,774-square-foot facility offers six pro-tournament courts with championship-broadcast lighting, locker rooms, two sports lounges, a 200-foot long glass-paneled observation loggia, two 120-foot long viewing balconies and an after-school homework area accessible via elevator or glass-tower stairway

Three primary challenges influenced the facility's design. The first challenge was housing six championship-sized tennis courts within a structure measuring 258 feet long by 175 feet wide by 55 feet high on a very tight site; zoning regulations required setbacks from the adjacent freeway and two side yards, and a mandatory utility setback from high-tension transmission lines. Secondly, the site had bay mud soil; the tennis courts required a slab that would not settle or crack so the solution involved poured-in-place and post-tensioned slabs set upon a system of grade beams and 186 auger-drilled friction piles set 70 feet into bay mud. The third challenge was cost. In order to economically construct a 175-foot clear-span space with 32-foot high court clearances, a modular pre-engineered steel-framed fabric-clad structure was developed and assembled "erector-set style" from shipping containers on site.

As envisioned by the owners—tennis professionals Anne and Horacio Matta—the new public arena provides the underserved and community-at-large with greater access to world-class tennis instruction.

Previous pages: Interior tennis courts
Above: Entry approach

Left: Night view of entry
Above: Lobby with view of courts

Left: View of courts from above
Above: Site and floor plan

227

30
ANNENBERG PROJECTS

30

安嫩伯格项目
ANNENBERG PROJECTS

ANNENBERG CENTER AT LOWER POINT VICENTE
RANCHO PALOS VERDES, CALIFORNIA
DESIGN 2009

BALLONA URBAN ECOLOGY CENTER
MARINA DEL REY,
LOS ANGELES, CALIFORNIA
DESIGN 2012

WALLIS ANNENBERG PETSPACE
PLAYA VISTA,
LOS ANGELES, CALIFORNIA
DESIGN 2015

Over the past 10 years, Swatt | Miers has been blessed with the opportunity to design three related yet entirely different projects for the Annenberg Foundation, each within the greater Los Angeles area. The first is at Lower Point Vicente in Rancho Palos Verdes; the second is within the Ballona Wetlands Ecological Reserve near Marina Del Rey; and the third is in Playa Vista, just east of the Ballona Reserve and among the emerging Los Angeles tech industry. The stories of these three amazing projects are woven together into a unique mosaic that is simultaneously joyous and bittersweet and certainly too long to adequately do justice within this publication. In summary, the first two, although extensively developed, were not constructed due to various political factors; the third is now under construction.

The first of these projects, the Annenberg Center at Lower Point Vicente, commenced in 2006 and featured a state-of-the-art animal care facility on an unbelievably wonderful site along the California coast. This world-class facility was to serve as a family-education destination, celebrating the peninsula's unique natural environment and the animals and other living organisms that have inhabited the site over time. Located on a magnificent bluff overlooking the Pacific Ocean, the Center complimented the existing Point Vicente Interpretive Center, which is a favorite destination for whale watchers from around the globe.

As part of the Annenberg Center, two new buildings, the Annenberg Companion Animal Center and the Outpost, a small arrival orientation facility, provided interactive educational exhibits, as well as volunteer and community service opportunities, animal behavioral training, adoption, and in-house veterinary services. Visually anchored to the land with cast-in-place concrete frames set on the site's uphill side, the Center's architecture has a strong horizontal thrust with deep cantilevered roof extensions, visually soaring towards the ocean and Catalina Island beyond. In order to preserve surrounding views, the building was excavated into the hill with the three-level structure gently stepping with the topography. Thus, when viewed from above from Palos Verdes Scenic Drive, the green-roofed building appears appears to rise up from the landscape. When viewed from below, the building's form is sculpted with three gentle arcs that allude to the shape of the shoreline.

The second commission, Ballona Urban Ecology Center, incorporated programmatic aspects of the initial project. However given the new site's unique ecological sensitivity within the Ballona Reserve, including the adjacent and controversial Ballona Wetland's restoration effort, Annenberg's visionary Executive Director, Leonard Aube, saw an opportunity to integrate philanthropist Wallis Annenberg's goals for domestic animals into a larger urban paradigm embracing the ecology of all living organisms within the Los Angeles metropolitan area. Water was the central chord that would weave all these urban components together.

Thus, the design for this facility focused on allowing the building itself to reflect the program's sustainable urban ecology principles through the development of:

- a net-zero energy system comprised of passive-energy design principles, including natural ventilation, solar mass, and high-efficiency solar and fuel-cell power sources

- a net-zero habitat-loss building design achieved through the introduction of extensive green roofs and by raising portions of the building above the ground so as to let the natural habitat run below

- water conservation systems and programs, including rainwater harvesting, waste-water, recycling and the use of low water-usage appliances and cleaning systems

- an indoor-outdoor environment that promoted both energy efficiency and good health (fresh air) while reflecting the unique nature of the Los Angeles climate

- natural building materials, such as architectural concrete and wood in the main building. Concrete assists the building's solar-mass capability and minimizes long-term maintenance, while wood accents from sustainable forests serve as sun screens and provide a sense of warmth and welcome

- habitat sensitivity to mirror the Center's interior exhibit's focus on native wildlife and habitat; the building's exterior features green roofs designed to help serve as "pollinators" for local wildlife, such as butterflies, while incorporating state-of-the-art techniques, such as the use of bird safety glass and related design features, to protects birds in flight.

The third project, now under construction, is the Wallis Annenberg PetSpace. Unlike the first two projects that featured all new construction and a strong Swatt | Miers indoor-outdoor design, PetSpace occurs in one of five existing buildings in the Playa Vista commercial center. Feeling that 10 years was long enough to wait for the animal care component of this project to be realized, the decision was made to enter into a 12-year lease and construct a state-of-the-art public-oriented facility. The 30,000-square-foot facility is due to open in spring of 2017.

Previous pages: Annenberg Center project viewed from entrance portal
Above: Annenberg Center project viewed from the west

Top: Annenberg Center project engaging the hillside, visually pointing to the Pacific Ocean
Bottom: View of Outpost exhibit building
Opposite: Annenberg Center project site plan

1. Outpost
2. Annenberg Companion Animal Center
3. Existing Point Vicente Interpretive Center

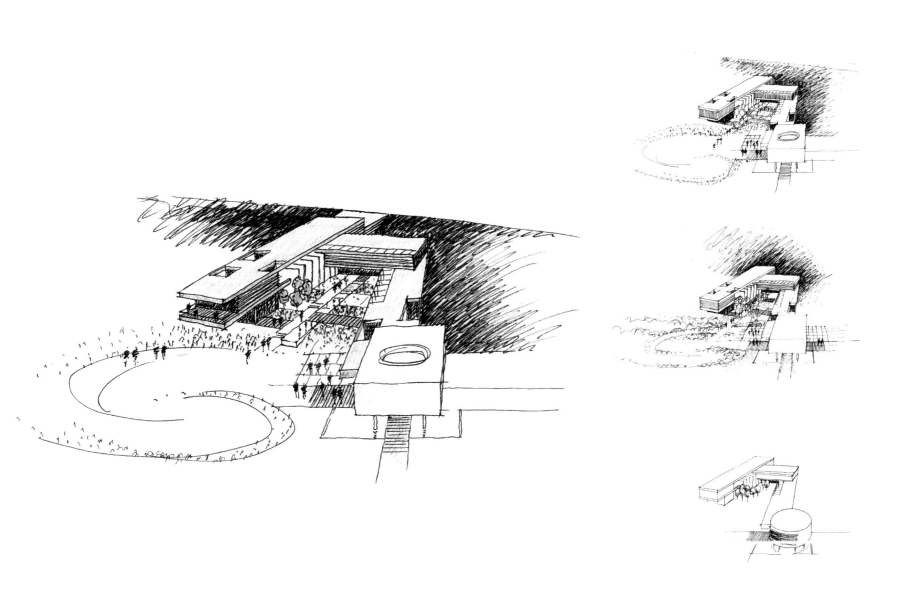

Opposite top: Courtyard entrance of Ballona Urban Ecology Center
Opposite bottom: Entrance lobby of Ballona Urban Ecology Center
Above: Early concept sketches of Ballona Urban Ecology Center

Left: Aerial view of Ballona Urban Ecology Center
Top: Site area flanked by Jefferson Boulevard to the left and the Marina Freeway to the right
Bottom: View of Ballona Urban Ecology Center from on-ramp of Marina Freeway

项目版权信息
PROJECT CREDITS

01 OZ House
Project team: Robert Swatt, FAIA; Steven Stept, AIA; Ivan Olds; Kimie Jurado; Jeanie Fan; Leila Bijan; Julie Liberman, AIA; Shira Czarny; Bernard Liwanag
Consultants: Yu Strandberg Engineering, structural; Lea & Braze Engineering, civil; Ron Herman, landscape; Architectural Lighting Design, lighting; Miller Design Company, interiors; Ultimate Control, audio visual; Russ Berger Design Group, media room
Contractor: Lencioni Construction

02 Sinbad Creek Residence
Project team: Robert Swatt, FAIA; Jeanie Fan; Connie Wong
Consultants: Yu Strandberg Engineering, structural; Building Performance Services, green consultant;
Contractor: K Daniel Kenny & Company

03 Simpatico Prefab Prototype
Project team: Robert Swatt, FAIA; Ivan Olds; Steven Stept, AIA; Jeanie Fan; Leila Bijan
Consultants: Innovative Structural Engineering, structural; Huettl Landscape Architecture, landscape; Monterey Energy Group, mechanical
Contractor: Eco Offsite

04 ARA House
Project team: Robert Swatt, FAIA; Steven Stept, AIA; Kimie Jurado; Ivan Olds; Leila Bijan; Jeanie Fan; Lauren Wynveen; Audrey Hitchcock, RIBA; Bernard Liwanag
Consultants: Yu Strandberg Engineering, structural; SWA, landscape; Lea & Braze Engineering, civil
Contractor: Lencioni Construction

05 Stein House
Project team: Robert Swatt, FAIA; Steven Stept, AIA; Ivan Olds, Leila Bijan, Lauren Wynveen, Jeanie Fan, Bernard Liwanag
Consultants: Yu Strandberg Engineering, structural; Thuilot Associates, landscape; Monterey Energy Group, mechanical; Neil O. Anderson & Associates, civil
Contractor: W. B. Elmer & Company

06 Vidalakis House
Project team: Robert Swatt, FAIA; Miya Muraki, AIA; Ivan Olds; Mike Eggers
Consultants: Yu Strandberg Engineering, structural; Lea & Braze Engineering, civil; Creekside Landscaping, landscape
Contractor: Lencioni Construction

07 Retrospect Vineyards House
Project team: Robert Swatt, FAIA; Julie Liberman, AIA; Audrey Hitchcock, RIBA; Connie Wong; Ivan Olds
Consultants: Yu Strandberg Engineering, structural; Bernard Trainor + Associates, landscape; Oberkamper & Associates, civil; Connie Wong, interiors
Contractor: Jamba Construction

08 Rashid House
Project team: Robert Swatt, FAIA; Steven Stept, AIA; Kimie Jurado; Ivan Olds; Brian Selkow; Mike Eggers
Consultants: Yu Strandberg Engineering, structural; Green Works, landscape; Lea & Braze Engineering, civil; Pritchard Peck Lighting, lighting
Contractor: Denis Matson

09 Oak Knoll House
Project team: Robert Swatt, FAIA; George Miers, AIA; Ivan Olds; Jeff Wheeler; Zachary Gong; Lauren Wynveen; Brian Selkow; Mili Del Castillo; Miya Muraki, AIA
Consultants: Yu Strandberg Engineering, structural; Lea & Braze Engineering, civil; Thuilot Associates, landscape; Monterey Energy Group, mechanical; Margraf Systems, audio visual
Contractor: CR Buildworks

10 Cheng-Brier House
Project team: Robert Swatt, FAIA; Michael Lehmberg, AIA; Miya Muraki, AIA; Connie Wong; Kimie Jurado; Vince Robles; Mili Del Castillo; Jennifer Kerrick; Phoebe Wong-Oliveros, AIA
Consultants: YU Structural Engineers, structural; Huettl Landscape Architecture, landscape; Lea & Braze Engineering, civil
Contractor: Jamba Construction

11 Kuenster-Miers House
Project team: George Miers, AIA; Matthew Lawhead
Consultants: Yu Strandberg Engineering, structural; Hogan Land Services, septic; PJC & Associates, geotechnical; Lea & Braze Engineering, civil; Mario Bazan, vineyard
Contractor: Paul Farinato Construction

12 Cheng-Reinganum House
Project team: Robert Swatt, FAIA; Julie Liberman, AIA; Misa Grannis; Connie Wong; Michael Lehmberg, AIA
Consultants: YU Structural Engineers, structural; Huettl Landscape Architecture, landscape
Contractor: W. B. Elmer & Company

13 Mora Estates
Project team: Robert Swatt, FAIA; Julie Liberman, AIA; Miya Muraki, AIA; Ivan Olds; Audrey Hitchcock, RIBA; Connie Wong, interiors
Consultants: Westfall Engineers, Inc., structural; Huettl Landscape Architecture, landscape; SMP Engineers, Civil; DZINE, furnishings and art
Contractor: Melvin Vaughn, founder, Vaughn-House

14 House in Northern India
Project team: Robert Swatt, FAIA; Steven Stept, AIA; Julie Liberman, AIA; Audrey Hitchcock, RIBA; Kimie Jurado; Jeff Wheeler; Miya Muraki, AIA;
Consultants: Urban Architecture Works, architect of record; Yu Strandberg Engineering, structural; Vintech Consultants, structural engineer of record; Shaheer Associates, landscape; Russ Berger Design Group, theater designer; Metro Eighteen, audio visual

15 Casa Santiago
Project team: Robert Swatt, FAIA; Ivan Olds; Kimie Jurado; Mathew Lawhead; Vince Robles; Kae Anantawong
Consultants: YU Structural Engineers, Structural; Blasen Landscape Architecture, Landscape; Lea & Braze Engineering, civil; Eric Johnson Associates, lighting; Jorie Clark Design, interiors; Russ Berger Design Group, acoustical; Monterey Energy Group, mechanical; Marchand Wright, audio-visual consultant

16 Helios House
Project team: Robert Swatt, FAIA; Julie Liberman, AIA; Connie Wong; Kimie Jurado; Leila Ghazavi; Ronald Tallon
Consultants: YU Structural Engineers, structural; Huettl Landscape Architecture, landscape; RSA+, civil
Contractor: Jamba Construction

17 Amara House
Project team: Robert Swatt, FAIA; Miya Muraki, AIA; Nana Koami; Kae Anantawong; Connie Wong; Yuko Okamura; Pete Austin
Consultants: YU Structural Engineers, structural; Bernard Trainor + Associates, landscape; Lea & Braze Engineering, civil
Contractor: Lencioni Construction

18 Whispering Stones House
Project team: Robert Swatt, FAIA; Michael Lehmberg, AIA; Lars Nilsson, AIA; Connie Wong; Kimie Jurado; Kae Anantawong; Johanna Malmerberg; Vince Robles; Pete Austin
Consultants: YU Structural Engineers, structural; Atterbury & Associates, civil; Bernard Trainor + Associates, landscape; Banks|Ramos, lighting; Susanna Van Leuwen, electrical; Monterey Energy Group, energy; CTC Associates, mechanical
Contractor: Ryan Associates

19 Irie House
Project team: Robert Swatt, FAIA; Ivan Olds; Ronald Tallon
Project architect: SCT Estudion de Arquitectura
Consultants: YU Structural Engineers, structural; Thuilot Associates, landscape; Via, technology; Estop Estudios Topogracia S.A., surveyor; Connie Wong, interiors

20 Orciuoli House
Project team: Robert Swatt, FAIA; Julie Liberman, AIA; Erik Waterman, AIA; Misa Grannis
Consultants: YU Structural Engineers, structural; Lea & Braze Engineering, civil; Thuilot Associates, landscape

21 Westwind House
Project team: Robert Swatt, FAIA; Phoebe Wong-Oliveros, AIA; Julie Liberman, AIA; Kimie Jurado
Consultants: Lea & Braze Engineering, civil; Murray Engineers, geotechnical; Bernard Trainor + Associates, landscape; Kielty Arborist Services, arborist

22 Chalon Road House
Project team: George Miers, AIA; Michael Gale, AIA
Consultants: Simpson Gumpertz & Heger, structural; LC Engineering Group, Inc., civil; Calwest Geotechnical Consulting Engineers, geotechnical; Land Phases, Inc., geologist; Zack Freedman, landscape; Building Solutions Group Roger M. Nite, mechanical; Kim Cladas, lighting

23 Lee House
Project team: Robert Swatt, FAIA; Phoebe Wong-Oliveros, AIA; Nana Koami; Vince Robles
Consultants: Schell and Martin, Inc., civil; Arborlogic Consulting Arborists, arborist; Huettl Landscape Architecture, landscape; Geotechnia Consulting Geotechnical Engineers, geotechnical

24 River House
Project team: Robert Swatt, FAIA; Ivan Olds; Cordelia Kotin
Consultants: Sol Development Associates, LLC, Entitlement; Bernard Trainor + Associates, landscape; Merchand Wright & Associates, technology; Specification West, specifications; YU Structural Engineers, structural; Dale G. Mell & Associates, surveyor; Monterey Energy Group, Inc., mechanical; Neumann Sloat Blanco Architects LLP, building envelope

25 Black Point House
Project team: Robert Swatt, FAIA; Julie Liberman, AIA; Erik Waterman, AIA; Leila Ghazavi
Consultants: YU Structural Engineers, structural

26 Kailua Beach Houses
Project team: Robert Swatt, FAIA; Julie Liberman, AIA; Connie Wong; Misa Grannis; Leila Ghazavi; Ronald Tallon
Consultants: YU Structural Engineers, structural; Huettl Landscape Architecture, landscape

27 SANA Residence
Project team: Robert Swatt, FAIA; Tricia Alesii

28 Palm Springs Animal Shelter
Project team: George Miers, AIA; Tim Hotz, AIA; Aaron Harte, AIA, LEED AP
Associated architect: Cioffi Architect
Consultants: KSP Consulting Engineers, Inc., structural; Merrick & Associates, mechanical; Randy Purnel Landscape Architects, landscape; Kruse & Associates, electrical; KEMA Services Inc., green building; Maureen Cornwell, interior design
Contractor: W.E. O'Neil Construction

29 Broadway Tennis Center
Project team: George Miers, AIA; Jeff Wheeler; Kae Anantawong
Consultants: John M Ward & Associates, planning; Simpson Gumpertz & Heger, structural; Fukushima Landscape Architecture, landscape; Lea & Braze Engineering, civil; Stanton Engineering, mechanical, electrical, plumbing; Global Fabric Structures, tennis court building envelope; Cornerstone Earth Group, geotechnical; Richard Sinner, specification writer; Kim Cladas, lighting
Contractor: W. L. Butler

30 Annenberg Projects
Client: The Annenberg Foundation and Wallis Annenberg
Executive Director: Leonard Aube; Cinny Kennard
Owner's Project Manager: Howard Litwak
Construction Manager: Patrick Ganahl
Senior Manager, Advocacy, Outreach, and Programs Advisor: Jackie Jaakola
General Manager: Carol Laumen
Exhibit Designer: Storyline

Annenberg Center at Lower Point Vicente
Architecture team: George Miers, AIA; Robert Swatt, FAIA; Colin Hlasny; Crystal DeCastro; Julie Liberman, AIA; Michael Reuter; Jan Leite; Jeanie Fan; Maureen Cornwell
Consultants: Simpson Gumpertz & Heger, Structural; WSP Flack & Kurtz, mechanical, electrical, plumbing; Melendrez, Landscape; KPFF Consulting Engineers, civil

Ballona Urban Ecology Center
Architecture team: George Miers, AIA; Robert Swatt, FAIA; Erik Waterman, AIA
Consultants: Simpson Gumpertz & Heger, structural; KPFF Consulting Engineers, civil; WSP Built Ecology, mechanical, electrical, plumbing; Melendrez, landscape

Wallis Annenberg PetSpace
Architecture team: George Miers, AIA; Robert Swatt, FAIA; Lars Nilsson, AIA; Erik Waterman, AIA; Angela Gormas; Maureen Cornwell
Consultants: John Labib + Associates, structural; ARC Engineering, mechanical, electrical, plumbing; PSOMAS, Civil; The Office of James Burnett, landscape; Banks|Ramos, Lighting; Edgett Williams Consulting Group, elevator; BBI Engineering, Inc., audio visual, low-voltage; Thornton Tomasetti, sustainability; Geoscience Analytical, Inc., methane; Richard Sinner, specification writer
Contractor: Matt Construction

当前团队
CURRENT & RECENT STAFF

Principals
Back row from left to right:
Phoebe Wong-Oliveros, AIA;
Michael Gale, AIA; Ivan Olds;
Julie Liberman, AIA; Miya Muraki, AIA.
Front row from left to right:
George Miers, AIA; Robert Swatt, FAIA.

Current Staff
Tricia Alesii
Behnaz Banishahabadi
Pete Austin
Eungoo Chong
Alberta Fujihara
Leila Ghazavi
Angela Gormas
Cheryl Hale
Kimie Jurado
Harlan Krusemark
Nana Koami
Lars Nilsson
Betty Nip
Yuko Okamura
Vince Robles
Sarah Sapone
Lorna Saunders
Marco Valgonio
Manuel Vivar
Connie Wong

Recent Staff
Kae Anantawong
Leila Bijan
Shira Czarny
Crystal DeCastro
Mili Del Castillo
Fanzheng Dong
Patti Donzelli
Mike Eggers
Jeanie Fan
Zachary Gong
Misa Grannis
Aaron Harte, AIA, LEED AP
Audrey Hitchcock, RIBA
Tim Hotz, AIA
Colin Hlasny
Niraj Kapadia
Jennifer Kerrick
Cordelia Kotin
Matthew Lawhead
Michael Lehmberg, AIA, LEED AP

Jan Leite
Bernard Liwanag
Johanna Malmerberg
Michael Reuter
Brian Selkow
Claire Sheridan
Steven Stept, AIA
Ronald Tallon
Erik Waterman, AIA
Jeff Wheeler
Lauren Wynveen
Jodie Zhang

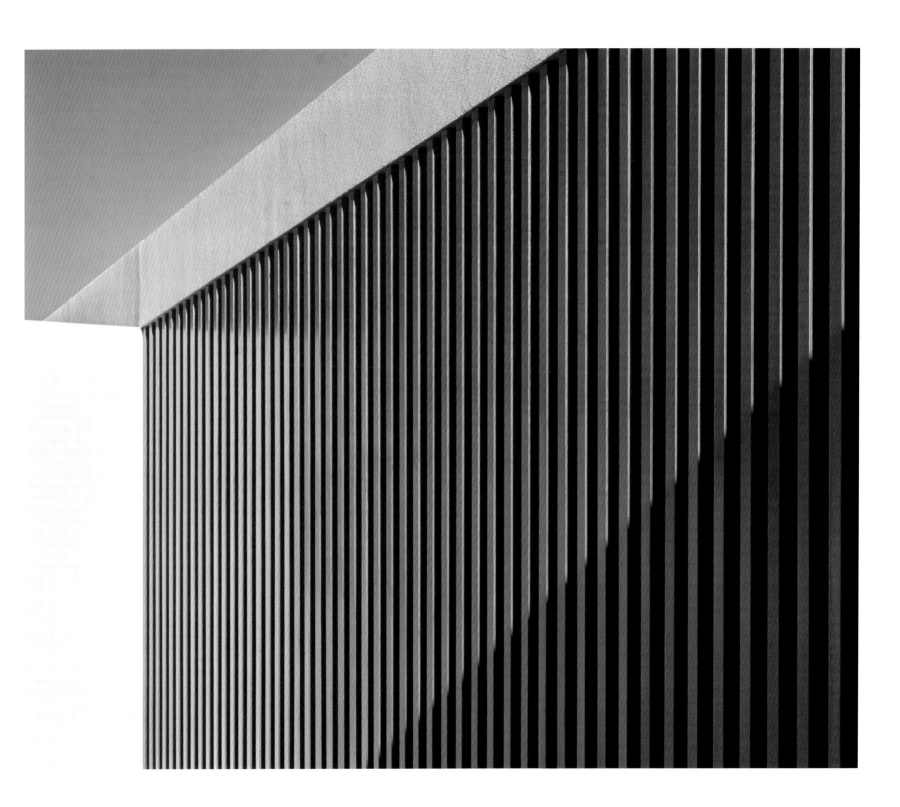

Above: Cheng-Brier House detail

人物介绍
BIOGRAPHIES

Robert Swatt, FAIA

Born in 1947, Robert Swatt grew up in Los Angeles where he was exposed to the works of California's early modern masters and gained an appreciation of architecture at a young age. He received his education at the University of California, Berkeley (UC Berkeley), and graduated with honors in 1970. Robert Swatt opened his own firm, Robert Swatt Architect, in 1975. He was a principal of Swatt & Stein Architects from 1977 to 1984 when he founded Swatt Architects in San Francisco.

Prior to starting his own firm, Robert worked with Howard A. Friedman in San Francisco and Cesar Pelli in Los Angeles. From 1975 to 1977 he taught architectural design at the University of California, Berkeley, where he has also served as a Director of the College of Environmental Design Alumni Association and was a founding member of the Distinguished Alumni Awards committee. He has lectured at the San Francisco Museum of Modern Art, Monterey Design Conference, Dwell on Design Conference, West Coast Green, and American Institute of Architects (AIA) National Convention, and exhibited work at the Monterey Design Conference, Modernbook | Gallery 494, LIMN, San Francisco AIA, California, UC Berkerley, and Cooper Hewitt, Smithsonian Design Museum in New York. In 1992 Robert was elected to the AIA College of Fellows, and in 2005 he was named a Howard A. Friedman Visiting Professor of Architecture at UC Berkeley.

In 2009, Swatt Architects merged with George Miers & Associates, becoming Swatt | Miers Architects, based in Emeryville, California. The firm has built an international reputation for design excellence covering an extraordinary variety of project types. Recent projects include award-winning and internationally acclaimed private residences throughout California and in Hawaii, Colorado, Canada, India, and Spain. Swatt | Miers has been recognized with more than 75 design awards, including 12 from the AIA; a National Honor Award for The Icehouse, Levi Strauss & Co. Corporate Headquarters; Residential Architect's 201 Project of the Year for the Tea Houses; International Architecture Award for OZ House and American Architecture Awards for Vidalakis House and Retrospect Vineyards House from the Chicago Athenaeum: Museum of Architecture and Design.

Articles on Robert's work have been published in in *Architecture*, *Architectural Record*, *Domus*, *GA Houses*, *Dwell*, *Hinge*, *Interior Design*, *The New York Times*, *Process Architecture*, *Progressive Architecture*, *Wallpaper*, and other journals and magazines in the United States and abroad. Private residences have been featured in numerous books published in the United States, Japan, China, Germany, United Kingdom, and Australia.

George Miers, AIA

Born in Fort Worth, Texas, in 1949 and raised in San Francisco, George Miers received his undergraduate education at the School of Architecture at Washington University in St. Louis. In addition to being a founding member of the United State's first Social Work and Architecture dual-degree program, George had the opportunity to study under a wonderfully diverse group of architects including Charles Moore, Ricardo Legorreta, and Dolf Schnebli.

After completing his Master of Architecture at UC, Berkeley, George began practicing in San Francisco during the mid-1970s. He worked as a designer on a wide range of large-scale special-use building types at Kaplan Mclaughlin Diaz and Skidmore, Owings & Merrill. During this time frame California's recently designated State Architect, Sim Van der Ryn, asked him to join a newly formed architectural programming unit whose task it was to set the standard for all future state buildings. In the course of setting these standards, George wrote the guidelines and user program for the State of Caliornia's first Energy Efficient Office Building in 1976, which launched a series of energy-efficient building designs throughout the state. The program for these facilities, while focused on sustainable architecture, placed an emphasis on balancing user comfort and operational needs within the framework of a strong architectural design. This directive—to integrate strong architectural principles with user and operational needs in a meaningful manner that facilitates the user's programs—formed the underlying design approach of George Miers and Associates, which he opened in San Francisco in 1982.

During the firm's 27 years of practice, prior to merging with Swatt Architects, George designed a wide range of award-winning public service buildings, including police facilities, civic centers, libraries, and animal care facilities. Many of these, such as the Antioch and Lodi police facilities and the San Diego Campus for Animal Care, have set the standard for these building types throughout North America.

部分年表
SELECTED CHRONOLOGY

Robert Swatt Architect

1975

Swatt/Luckham/Bennett (Amito I) House, Berkeley, California

1976

Swatt/Everts (Amito II) House, Berkeley, California

Swatt & Stein Architects

1976

Paganelli House, Oakland, California

1977

Overlook Houses, Walnut Creek, California

Swahlen House Addition, San Rafael, California

Competition for an Energy Efficient Office Building, Sacramento, California

Brodsky House, Scotts Valley, California

Foster/Feldman House, Berkeley, California

1978

Conroy House Terraces, Berkeley, California

Crocker Bank Retail Banking Headquarters, San Francisco, California

Sommers House, Sausalito, California

Weissberg House, Oakland, California

1979

Wrubel House Addition, Berkeley, California

Riback/Steffen House, Oakland, California

Carico House, San Francisco, California

Taunton House, Lake Tahoe, California

Ionic Building, Oakland, California

Laio House, Berkeley, California

1980

New Entrance to the Japanese Tea Garden, Golden Gate Park, San Francisco, California

H.I.S. Building, Japantown, San Francisco, California

Miller House, Santa Cruz, California

Mirabeau Greenhouse Restaurant, Oakland, California

1981

Crocker Banks, Salinas and Sacramento, California

Prusinski House Addition, Orinda, California

Barratt Studio Solo, Sacramento, California

Okabe Sports, San Francisco, California

Iroha Restaurant, San Francisco, California

1982

Gerson Bakar Greenhouse, San Francisco, California

Barratt Infill Housing, Sacramento and Emeryville, California

Heilbron Square, Sacramento, California

Bank of America, San Francisco International Airport, San Francisco, California

1983

Bill Graham House Addition, Corte Madera, California

Swatt Architects

1984

Wilsey Conservatory, San Francisco, California

Koret Manufacturing Plant, Price, Utah

Treat Executive Center Interiors, Walnut Creek, California

Koshland House, Berkeley, California

1985

Herrick/Kunitz House, Berkeley, California

UC Berkeley Business School, Feasibility Study and Schematic Design, Berkeley, California

Bauch House, Belvedere, California

Osgood/Coppola House, San Francisco, California

Kingsley House Addition, San Francisco, California

1986

Thom House, Woodside, California

Strunsky House Remodel, San Francisco, California

Levi Strauss & Co. Finishing Center, San Francisco, California

1987

Treble House Addition, Palo Alto, California

Jewish Home for the Aged Boutique, San Francisco, California

Levi Strauss & Co. Regional Sales Offices, Atlanta, Georgia

Schrag/Marinoff Studio, Berkeley, California

Levi Strauss & Co. Finishing Center, San Antonio, Texas

1988

Friedman House Addition, San Francisco, California

Plaza Café Remodel, Carmel, California

American Tin Cannery Master Plan, Pacific Grove, California

Levi Strauss & Co. Distribution Shipping Area, Henderson, Nevada

Levi Strauss & Co. Conference Center, San Francisco, California

Abraham House, Orinda, California

Levi Strauss & Co. Master Plan, San Francisco, California

Bauch Beach House, Stinson Beach, California

1989
Goldman House Addition, San Francisco, California

Esther & Jacques Reutlinger Community for Jewish Living, Danville, California

Levi Strauss & Co. History Museum, San Francisco, California

1990
The Icehouse, Levi Strauss & Co. Headquarters, San Francisco, California

Haas Ranch House Remodel, Big Timber, Montana

Haas House Solarium, San Francisco, California

Goldman Vacation House, Lake Tahoe, California

Levi Strauss & Co. Showrooms and Regional Sales Offices, Los Angeles, California

1991
Levi Strauss & Co. Showrooms and Regional Sales Offices, San Francisco, California

Levi Strauss & Co. East Bay Offices, Walnut Creek, California

Blatteis House Addition, Oakland, California

Adcock House, Oakland, California

Barney House, Oakland, California

Levi Strauss & Co. Data Center, San Francisco, California

Swatt House Remodel, Oakland, California

1992
Richard & Rhoda Goldman Fund Offices, San Francisco, California

Peter & Mimi Haas Fund Offices, San Francisco, California

Evelyn & Walter Haas Jr. Fund Offices, San Francisco, California

Columbia Foundation Offices, San Francisco, California

Larsen/Wong House, Oakland, California

Icehouse Alley, San Francisco, California

1993
Wells Fargo Branch Bank, Fresno, California

544 2nd Street Building Alterations, San Francisco, California

Koshland House Addition, Lafayette, California

UC Berkeley Evans Hall Classrooms, Berkeley, California

1994
BR Cohn Winery Hospitality Center Study, Glen Ellen, California

Ocean View House, Oakland, California

Growers Square Renovations, Walnut Creek, California

Stanford University Dormitories, Stanford, California

Mikimoto Boutique, San Francisco, California

Abbey House Alterations, San Francisco, California

UC Berkeley Telecommunications Offices, Berkeley, California

Dominican College Classroom Standards, San Rafael, California

1995
Swatt House, Lafayette, California

Levi Strauss & Co. Saddleman Building Renovation, San Francisco, California

Glen Ellen Winery Tasting Room and History Center, Glen Ellen, California

Fritzi California Master Plan, San Francisco, California

Pitney Bowes Offices, San Francisco, California

Chetkowski House Addition, Piedmont, California

UC Berkeley Memorial Stadium Club Room, Berkeley, California

1996
Levi Strauss & Co. Showrooms and Regional Sales Offices, San Francisco, California

UC Davis Laboratory Alterations, Davis, California

1997
UC Davis Recreation Hall Upgrades, Davis, California

UC Davis Chancellor's Conference Room, Davis, California

Kohavi/Lace House, Portola Valley, California

Wellington Vineyards Tasting Room, Glen Ellen, California

Stark/Drayer House Addition, Oakland, California

1998
Meridian Winery Exhibit, Paso Robles, California

UC Berkeley Free Speech Movement Café, Berkeley, California

UC Berkeley Microcomputer Center, Berkeley, California

Palo Alto House, Palo Alto, California

Codorniu-Napa Carneros Museum, Napa, California

Wildhourse Winery Tasting Room and Hospitality Center Design, Templeton, California

Crossbrook Drive Homes, Moraga, California

Woodland House, Kentfield, California

1999
Mayol House, Escalon, California

UC Berkeley Media Resources Center, Berkeley, California

St. Marks Episcopal Church Restoration, Berkeley, California

Rosen House Remodel, Lafayette, California

UC Davis Bainer Hall Laboratory Renovations, Davis, California

Kamat House Addition and Remodel, San Carlos, California

Greason House Addition, Lafayette, California

UC Davis Mrak Hall Administration Building Interiors, Davis, California

Green City Lofts, Emeryville, California

Nelson Communications, Sacramento, California

Vignos House Addition, Orinda, California

Hoffman Houses, Lafayette, California

Conrad House, Sausalito, California

2000
Lea House Addition, Orinda, California

CBS Marketwatch.com, San Francisco, California

Shimmon House I, Los Altos Hills, California

Hoffman Secluded Place House, Lafayette, California

2001

Palo Alto Guest House, Palo Alto, California

Gradman House, Inverness, California

2002

Sackett-Mohsenin House, Ross, California

Shimmon House II, Los Altos Hills, California

St. John's Episcopal Church School, Ross, California

Orr House, Saratoga, California

Tea Houses, Silicon Valley, California

2003

Lukaszewicz-Moayeri House Addition, Hillsborough, California

Chang-Worzel House Remodel, San Francisco, California

Morley House, Lafayette, California

King House Addition and Pool House, Lafayette, California

2004

Leung-Chen House, Hillsborough, California

Lo House Remodel and Addition, Palo Alto, California

Ting-Ho House Remodel and Addition, Hillsborough, California

Binger House, Oakland, California

Powell-Sukys House Remodel and Addition, Orinda, California

2005

Methven House, San Anselmo, California

Garay House, Tiburon, California

Four-Generation House, Larkspur, California

Tsou House Addition and Remodel, San Carlos, California

Kapoor House, Berkeley, California

2510 Skyfarm House Addition and Remodel, Hillsborough, California

2006

Elizabeth House, Walnut Creek, California

Seneca Center, San Leandro, California

OZ House, Atherton, California

Hudson-Panos House, Healdsburg, California

2007

Orr Deck and Exercise Pavilion, Saratoga, California

Rittenhouse Avenue House Remodel, Atherton, California

Lub House, Napa, California

Macomber House, Oakland, California

Eliasaf-Shoham House Remodel, Los Altos, California

Braz-Pereira House Remodel, San Rafael, California

Sugarman House Addition and Remodel, Santa Rosa, California

Simpatico Prefab Prototype, Emeryville, California

Elizabeth House II, Walnut Creek, California

Glazer-Morton House, Point Reyes, California

2008

Stein House, Orinda, California

Peninsula Humane Society/SPCA, Burlingame, California

Sinbad Creek House, Sunol, California

Hill Way House, Los Altos Hills, California

Portofino Riviera Conceptual Design Study, Sausalito, California

Glass House Addition and Remodel, Alamo, California

Rosa-Poffenbarger Loft Remodel, San Francisco, California

Kimball Pool House, Kentfield, California

Koshland House Remodel, Berkeley, California

Baru House, Oakland, California

Gibbs House Remodel, San Francisco, California

Bewsher House, Tiburon, California

Ara House, Atherton, California

Swatt | Miers Architects

2009

The Annenberg Project at Lower Point Vicente, Palos Verdes, California

Annenberg Outpost, Palos Verdes, California

Annenberg Rancho Palos Verdes, California

City of Palm Springs Animal Shelter, Palm Springs, California

Sonoma County Animal Shelter, Santa Rosa, California

Sutter County Animal Shelter, Yuba City, California

Merced County Emergency Operations Center, Merced, California

Ottawa Humane Society, Ottawa, Ontario, Canada

Regina Humane Society, Regina, Saskatchewan, Canada

Parisi-Dunne House Addition and Remodel, Pacifica, California

Vidalakis House, Portola Valley, California

2010

Attestatova-Laliberte House Remodel and Addition, Orinda, California

Lang-Jiang House, Tiburon, California

City of Clovis Animal Shelter, Clovis, California

Lederman House Addition and Remodel, San Rafael, California

Retrospect Vineyards House, Windsor, California

Hagar Avenue House, Piedmont, California

Schwinn-Cherry Prefab House, Orinda, California

Kirby House, Gibsons, British Columbia, Canada

Foley House, Honolulu, Hawaii

Santa Cruz SPCA, Santa Cruz, California

Rashid House, Los Altos, California

Kobs Prefab House, Alamo, California

Lloydminster SPCA, Lloydminster, Saskatchewan, Canada

2011

Bhatia Addition, Piedmont, California
Barlow Addition and Remodel, Lafayette, California
Mora Estates, Lot 2, Los Altos Hills, California
Lau Houses, El Cerrito, California
Placer SPCA, Roseville, California
Oh House Remodel, Lafayette, California
Morrison House, San Anselmo, California
East Bay SPCA, Oakland, California
Wong-Huh Addition and Remodel, Tiburon, California
Epstein Addition and Remodel, Orinda, California
Yiu House, Los Altos Hills, California
House in Northern India, New Delhi, India
Cheng-Brier House, Tiburon, California
Oak Knoll House, Napa, California

2012

Medicine Hat SPCA, Medicine Hat, Alberta, Canada
Ballona Urban Ecology Center, Marina Del Rey, Los Angeles, California
Evans Remodel, Los Altos Hills, California
Ledesma Remodel, Orinda, California
Orr Exercise Pavilion, Saratoga, California
OZ House Accessory Structures, Atherton, California
Keunster-Miers House, Sonoma, California
Bakersfield Animal Shelter, Bakersfield, California
Capper Felmly Music Room Addition, Walnut Creek, California
Yoon House, Palo Alto, California
Pets Lifeline, Sonoma, California
Vaughn Houses, Los Altos Hills, California
Broadway Tennis Center, Burlingame, California

2013

Orciuoli House, Atherton, California
Medeiros House, Orinda California
Cheng-Reinganum House, Orinda, California
Canyon View House, Los Angeles, California
Kunkel Remodel, Portola Valley, California
Casa Santiago, Atherton, California
Kern County Animal Shelter, Bakersfield, California
Sacks House, Mill Valley, California
Helios House, Napa, California
Winters Master Plan, Orinda, California
Grant House, Tiburon, California
Lafayette Community Center Design, Lafayette, California
Kim House, Los Altos Hills, California
Dal Bozzo House, Ross, California
Wilder House I, Orinda, California
Vineis House, Napa, California

2014

Whispering Stones House, Healdsburg, California
Wilder House II, Orinda, California
Amara House, Atherton, California
Brandt House, Tiburon, California
Cinnamon Ventures House, Hillsborough, California
Mora Estates: Lot 3, Los Altos Hills, California
Duncan Street Remodel, San Francisco, California
S. Canyon View Drive House, Brentwood, California
Pacific Grove House, 17 Mile Drive, Pacific Grove, California
Irie House, Balearic Islands, Spain
Kailua Beach Houses, O'ahu, Hawaii
Matas Office Structure and Landscape, Healdsburg, California
Yandell House, San Anselmo, California
Duncan Street Guest House, San Francisco, California
Owyang Remodel, Atherton, California
Dodson-Jacobs House, Los Altos Hills, California
Bournemouth Road House, St. Helena, California
Westwind House, Silicon Valley, California
Grizzly Peak House, Berkeley, California
Chalon Road House, Bel Air, Los Angeles, California

2015

Wan House, Tiburon, California
Lee House, Sonoma, California
Wallis Annenberg PetSpace, Los Angeles, California
Katz House Addition, Sausalito, California
Toy Remodel, Woodside, California
Boladian Addition and Remodel, Hillsborough, California
Liu-Johnson House, Durango, Colorado
River House, Fresno, California
Mason House, Los Altos Hills, California
OZ House Addition, Atherton, California
RGen House, Atherton, California
Lee House, Lafayette, California
Roberts Residence: Carriger Road, Sonoma, California
Morris Prefab House, Los Angeles, California
Stewart House, Santa Rosa, California
Black Point House, Honolulu, Hawaii
Mora Estates Lot 1, Los Altos Hills, California

2016

Larry Walker & Associates, Davis, California
Pets Lifeline, Sonoma, California
Schrag Marinoff Addition and Remodel, Berkeley, California
Osterloh House, Hillsborough, California
SANA Residence, Los Altos Hills, California
Dragonfly House, Palo Alto, California
Evans Conceptual Design, Lafayette, California
Jamie Roberts Residence, San Francisco, California
McFadden Residence, Doheny Estates, Los Angeles, California
Buddhist Temple Housing Studies for Prometheus Real Estate, Mountain View, California
Black Mountain Housing for Dragonfly Group, San Carlos, California

出版物
PUBLICATIONS

Books

100 More of the World's Best Houses. Melbourne, Australia: Images Publishing Group, 2005.

1000 X Architecture of The Americas. Verlagshaus Braun: Deutsche Bibliothek, 2008.

Abraham, Russell. *California Cool: Residential Modernism Reborn*. Melbourne, Australia: Images Publishing, 2010.

---. *California Cool: Residential Modernism Reborn (New Edition)*. Melbourne, Australia: Images Publishing, 2013.

---. *Rural Modern*. Melbourne, Australia: Images Publishing, 2013.

Asenio, Francisco. *House in Sunshine – The North American Best Selection of Contemporary Architectural Houses*. Hong Kong: Pace Publishing Limited, 2007.

Beaver, Robyn. *21st Century Houses: 150 of the World's Best*. Melbourne, Australia: Images Publishing Group, 2010.

---. *Another 100 of the World's Best Houses*. Melbourne, Australia: Images Publishing Group, 2003.

---. *A Pocketful of Houses*. Melbourne, Australia: Images Publishing Group, 2006.

---. *The New 100 Houses x 100 Architects*. Melbourne, Australia: Images Publishing Group, 2007.

Binder, Georges, et al, ed. *International Architectural Yearbook: No.2*. Melbourne, Australia: Images Publishing Group, 1996.

Browne, Beth. *Masterpiece: Iconic Houses by Great Contemporary Architects*. Melbourne, Australia: Images Publishing, 2012.

City by Design: An Architectural Perspective of the Greater San Francisco Bay Area. Plano, Texas: Panache Partners, 2009.

Cleary, Mark. *200 Houses*. Australia: Images Publishing, 2011.

Cook, Jeffery. *Award Winning Passive Solar House Designs*. Charlotte, Vermont: Garden Way Publishing Co, 1984.

Dream Homes: Northern California. Plano, Texas: Panache Partners, 2007.

Fatih, Driss. *Furniture by Architects*. Victoria, Australia: Images Publishing, 2013.

---. *Pure Luxury: World's Best Houses*. Australia: Images Publishing, 2012.

GA House 64. Tokyo, Japan A.D.A Edita, March 2000.

GA House 84. Tokyo, Japan A.D.A Edita, July 2004.

Galindo, Michelle. *Masterpieces: Bungalow Architecture + Design*. Germany: Braun Publishing, 2013.

Health Spaces of the World: Volume 2. Melbourne, Australia: Images Publishing Group, 2003.

Home: New Directions in World Architecture and Design. Cincinnati, Ohio: How Books, F + W Publications, 2006.

Hu, Yanli. *Architecture Highlights 2*. Kowloon, Hong Kong: Shanglin A & C Limited, 2009.

Ichinowatari, Katsuhiko Abercombie, ed. *Process Architecture No. 18: Modern Wooden Houses*. Tokyo: Process Architecture Publishing, 1980.

Interior Spaces of the USA: Volume 3. Melbourne, Australia: Images Publishing Group, 1997.

Lam, George. *HA: House*. Hong Kong: Pace Publishing, 2010.

---. *House & Housing 101*. China: Pace Publishing Ltd, 2012.

Residential Spaces of the World: Volume 1. Melbourne, Australia: Images Publishing Group, 1994.

Residential Spaces of the World: Volume 2. Melbourne, Australia: Images Publishing Group, 1997.

Sardar, Zahid. *San Francisco Modern: Interiors, Architecture & Design*. San Francisco, California: Chronicle Books, 1998.

Seaside Living: 50 Remarkable Houses. Victoria, Australia: Images Publishing, 2015.

Swatt Architects: Livable Modern. Melbourne, Australia: Images Publishing Group, June 2004.

Swatt | Miers Architects. *Inside Out: New Modern West Coast Architecture*. Australia: Images Publishing, 2010.

Trulove, James Grayson. *Living Outside Inside*. Harper Design International, New York, June 2004.

---. *The New American House 4: Innovations in Residential Design and Construction*. New York: Whitney Library of Design, Watson-Guptil Publications, 2003.

Magazines & Periodicals

"12 Crown Jewels of Desert Modernism," *Desert Magazine*, February 2012.

"1983 Design Awards," *Architectural Record*, July 1983.

"2010 Builder's Choice," Garay House, Tea Houses. *Builder Magazine*, October 2010, 76, 96.

Acsay, Judit. "Körkörösen Nyitott," *Villák Magazine*, June 2005, 32–43.

"Adding High Light," *Sunset*, January 1991, 58–62.

"AEC Interview," Profile; *The Architects Sourcebook*, March 1999.

AIA East Bay Chapter. "Firm Profile: Swatt | Miers Architects," *ARCHnews*, July 2010, 8.

AIA East Bay. "House Matter," *ARCH News*, April 2006.

AIA East Bay. "Sinbad Creek Residence," *Diablo Magazine*, September 2013, 101–3.

Anderson, Judith. "The Architecture of Interior Design," *San Jose Mercury News*, April 3, 1988, 20–6.

Ben-Yehuda, Rina. "San Francisco on the Water," *ITSUV*, April 2006, 26–30.

Bergeron, Caroline. "A la Californienne – California Dreamin'," *Prestige Design* 10, March 2013, 36–42.

Bertleson, Ann. "Light Box," *Sunset*, October 1997, 118.

Bertleson, Ann. "Special Weapons and Tactics," *Northern California Home & Garden*, May 1990, 44–51.

Bertleson, Ann. "Stairways," *Northern California Home & Garden*, October 1992, 69, 120.

"Bold Composition," *myTrends Home*, November 22, 2015, 6–15.

Botello, Alfredo. "The Idea House," *Diablo Magazine*, March 1999, 38–55.

Brettkelly, Jody. "Couple create hillside haven at Sunol's Sinbad Creek," *San Francisco Chronicle*, March 28, 2014. Home & Garden.

Brown-Martin, Darcy. "The Mod Squad," *Diablo Magazine*, April 2002, 48–51.

"California Contextualism," *Architectural Record*, October 1983.

"California Cool: The architectural dream team of Robert Swatt and George Miers conceives a minimalist chic dream home in Tiburon," *Gentry Home*, March–April 2016, 38–45.

"California Dreaming," *TREND*, May/June 2012, 62–72.

"California Living: Swatt Miers Architects," *Diablo Magazine*, October 2016, 133.

Carlston, Lon M. "Enough of the Earthtones," *Oakland Tribune*, August 15, 1976.

Carvajal, Jose. "Lake Elsinore waffling on animal shelter," *The Californian*, February 23, 2007.

Chatfield-Taylor, Joan. "Flight of Fancy," *Design for Living*, Spring 2006.

Clark, Susannah. "Comfort and Joy," *Diablo Magazine*, December 1994, 36–8.

Coupland, Ken. "Bold Designs Forged From Fire," *San Francisco Examiner*, February 28, 1993, F1.

Crooks, Peter. "Dreamscape," *Diablo Magazine*, February 2011, 50–7.

Cuff, Denis. "New Shelter Gives Critter Better Odds," *Contra Costa Times*, May 17, 2005.

Despres, Delphine. "Demesure Californienne," *Artravel* 45, Summer Issue, 96–103.

Dorn, Suzanne. "Life Care Village Will Rise in Contra Costa," January/February 1993, 16.

Drueding, Meghan, Weber, Cheryl, Snider, Bruce D., and Maynard, Nigel, F. "Residential Architect Design Awards 2010," *Residential Architect*, May/June 2010, 32–5.

Drueding, Meghan. "Sweet and Low-key- Five Step Plan," *Custom Home*, July/August 2007, 90–2.

Eng, Rick. "Permanent Fashion," *Designers West*, November 1991, 80–3.

Fauntleroy, Gussie. "The Art of the House: A harmonious client-architect match produces a home with artistic touches at every turn in Portola Valley, California," *Western Art & Architecture*, June–July 2016, 116–21.

"Feeling Home Away from Home," *DO Magazine*, March/April 2007.

Feldman, Deborah. "Ironic Ionic," *Domus*, August 1980.

Fernandes, Anais. "Corredor de Ar," *Folho de Sao Paolo*, November 2014, 2.

Ferrari, Massimo. "L'Oriente visto dalla California," *Casabella* 794, October 2010, 68–75.

Ferri, John. "1223 Upper Happy Valley Road Boasts Spectacular Vistas," *San Francisco Chronicle: Real Estate*, February 5, 2012, K4–5.

"First AIA Honor Awards for Interiors," *Progressive Architecture*, June 1994, 70.

Foote, Justin. "California Dreaming," *TRENDS* 27, no. 9, 6–14.

Fuhrman, Janice. "An Expansive Contemporary," *California Homes*, March/April 2002, 82–9.

Furio, Joan. "Modular Programming," *Dwell*, Dec/Jan 2013.

Furio, Joanne. "Healdsburg Home's Green Aspects Can't Be Seen," *San Francisco Chronicle*, February 28, 2010.

Furio, Joanne. "Highly Connected," *The Robb Report Collection*, September 2009, 60–6.

Garlock, Judy. "Designing with Glass," *Beautiful Homes*, Winter 2007, 114–19.

Gaspard Sanchez, Christine. "Esprit Cabane," *Artravel* 47, Fall Issue, 77.

"Going Green: Winnipeg Humane Society," *Wood Design & Building*, Spring 2008, 34–9.

"Gold Award For Historic Preservation," *San Francisco Focus*, July 1993.

Goode, Stephen. "Chairs Do More Than Sit Outdoors," *Insight*, July 4, 1988.

Goodwin Hemmings, Sonya. "Let the Sun Shine," *Silicon Valley Home*, May 2007, 25–7.

Gregory, Daniel. "Checkerboards of Glass," *Sunset*, February 1982.

Gregory, Daniel. "Over The Garage and Across the Living Room—A Bright New Balcony Deck," *Sunset*, January 1982.

Grossman, Daniel. "Neo-Classical (Again)," *Harpers & Queen*, December 1980.

Gundrum, Daniel. "Utilitarian Chairs Gone Wild and Witty are Abloom in Cooper-Hewitt's Garden," *New York Times*, Mary 19, 1988.

Haba, Péter. "Egymásba Olvadó Hagyományok," *Villák Magazine*, May 2005.

Haba, Péter. "Tiszta Tükörkép," *Villák Magazine*, April 2005, 36–41.

Harmanci, Reyhan. "A Green Approach," *Dwell*, December/January 2007, 95–8.

Harris, Sandra Ann. "Long, Lean, & Lovely," *California Home*, February 2008, 125–31.

Harris, Sandra Ann. "Modern in the Blood," *Design For Living*, Spring 2007, 63–6.

Harris, Sandra Ann. "Modern Moment," *Diablo Magazine*, April 2007, 76–83.

Hawkes, Colleen. "Another Plane," *TRENDS: Home & Architectural* 2701, Jan 2011, 48–51.

Igonda, Marcelo. "Tradicion Y Modernismo en Oakland, California," *La Opinion*, June 1983, 12–13.

"Interior Monologues," *Residential Architect*, January/February 2011.

Jayakar, Devyani. "Blurring the Boundaries," *Inside Out*, Issue 301, July 2010, 108–17.

Kalicanin, Jelena. "Samostalna Kuca," *Kuca Stil*, October 2010, 3–8.

Keates, Nancy. "Modern Among the McMansions," *Wall Street Journal*, January 21, 2011.

Kinzie, Pam. "Oakland / Berkeley Rebuild After 1991 Firestorm that Destroyed 2,800 Houses," *Architectural Record*, April 1993, 23.

Laidman, Dan. "Architecture Firm's Shelters Pamper Animals," *Contra Costa Times*, May 17, 2005, C1–2.

Lee, Lydia. "Building the Future: Village Green," *California Home & Design*, March 2007, 70–2.

Lee, Lydia. "Inside Out," *Modern Luxury Silicon Valley*, March–April 2016, 78–81.

Lobdell, Heather. "House Warming," *Better Homes and Gardens*, March 2009, 69–76.

"Long-term Care Facility," *Progressive Architecture*, August 1993, 27.

Lynch, Sarah. "The Way Home," *California Home & Design*, October 2006, 114–23.

MacIsaac, Heather Smith. "Ashes to Architecture," *House & Garden*, March 1993, 84.

Mack, Mark. "Small Spaces—Urban and Suburban Refinements," Fall 1980.

MacMasters, Dan. "A Bold Composition of Simple Forms," *Los Angeles Times Home Magazine*, March 5, 1978, 14–15.

Matteucci, Jeannie. "Opening Up a Ranch-Style Home," *San Francisco Chronicle*, July 18, 2010, L1.

Matteucci, Jeannie. "The Devil Is In The Details," *San Francisco Chronicle*, October 15, 2003, F1, 6.

Maynard, Nigel F. "GreenCity Lofts," *Builder,* October 2007.

McCrohan, Deirdre. "Ultramodern new home OK'd for Lyford Drive," *Ark Newspaper*, December 5, 2012.

Meckel, David. "Exhibit on Post Fire Houses in the Oakland Hills," *Progressive Architecture*, March 1993, 22.

Milshtein, Amy. "Fitting Rooms," *Contract Design*, June 1994, 70–3.

Mitchem, Scott. "Level Success," *Wallpaper*, October 2001, 199–200.

Moxam, Charles. "Far-sighted," *TRENDS: Home & Architectural* 2606, August 2010.

Moxham, Charles. "Exotic Welcome," *Home & Architectural Trends* 29, no. 2, February 2013, 18–25.

Mumford, Steve. "The Icehouse," *Buildings*, June 1993, 44–8.

Nash, Kay Chabot. "Kitchen Debates," *Diablo Magazine*, September 1999, 46–7.

"Objectif Nature: Les Maisons De Thé En Vogue À Silicon Valley," *Deco Magazine*, June–September 2013, 246–55.

O'hkura, Peggy. "Uniquely Designed Building Taking Shape on Buchanan Mall," *Hokubei Mainichi*, April 9, 1982.

"Outside In: Award-winning architect Robert Swatt, of Swatt Miers Architects, creates an instant classic in Orinda," *Gentry Home*, May/June 2013, 58–63.

Özer, Derya Nüket. "Palo Alto Evi," *Yapi*, February 2005, 60–3.

Patel, Nina. "Steel Away," *Remodeling*, September 2002, 58.

"Pet Project: The Greatest Animal Center in Known-History Comes to Silicon Valley," *The Wave*, March 2008, 12–25.

Porter, Anna Watkins. "Architectural Aspirations," *SPACES: Luxurious Living in the East Bay*, August 2006, 26–9.

Porter, Paige. "Poised to Soar," *New Home*, Spring/Summer 2008, 52–6.

Pritchett, Kathryn Loosli. "Outdoor Connection," *SPACES*, May 2010, 24–9.

Pui-Wing Tam and Nancy Keats. "Silicon Valley Reboots," *Wall Street Journal*, October 24, 2012.

"Purity of Line: Modern architecture perfected," *Gentry Home*, May–June 2016, 90–7.

Quintero Ovalle, Zandra. "El Abrigo del Bosque," *Revista Habitar*, March 2014.

"Rebirth of the Oakland Hills," *Sunset*, September 1994.

"Rural Home Seems to Sprout From the Landscape," *California Builder*, February/March 1991, 16.

Sakamoto, Timothy. "Planet Architecture: Bay Area Modern," *In-D Digital*, 2000.

Sardar, Zahid. "Light Meter: Robert Swatt Illuminates Dramatic Spaces in Palo Alto," *Western Interiors and Design*, September/October 2003: 74–83.

Sardar, Zahid. "Plane Living," *San Francisco Chronicle Magazine*, May 2007, 18–23.

Sardar, Zahid. "Plane Thinking," *San Francisco Examiner Magazine*, May 17, 1998, 76–9.

Sardar, Zahid. "Power Play," *San Francisco Examiner Magazine*, September 17, 2000, 14–19

Schleich, Jenn. "Scenic Vistas in Sunol," *S/Style Magazine*, Summer 2015, 104–7.

Smaus, Robert. "They Reached for the View," *Los Angeles Times Home Magazine*, November 1979.

Snir, Boaz. "Hotel California," *DO Magazine*, July/August 2006, 124–33.

Stinnard, Michelle. "Stone brings distinction to local library," *Stone World*, June 2005, 65–70.

"Swatt & Stein," *Architectura Madrid*, No. 220, September/October 1979, 36.

"Swatt Architects," *Small Firms, Great Projects*, Issue 2008/2009, 189.

"Swatt Architects: Icehouse Project," *Hinge* 2 August 1994, 36–7.

Taylor, Tracy. "Chasing the Light," *Diablo Magazine*, April 2014, 34–5.

"Tea Houses," *Konsept Projeler*, October 2011, 70–77.

"Tea Houses: Swatt | Miers Creates a Special Blend for California's Silicon Valley," *Malibu HOME Magazine*, January 2013, 102–6.

Tilton, Sarah. "Berkeley House Features Great Views, Modern Style," *Wall Street Journal*, February 23, 2012.

Tracy-Williams, Laura. "Puppy Love," *Green Building & Design*, July–September 2012, 165–7.

Tucker, John G. "Runway Visibility," *Interior Design*, April 1984.

Wagner, Michael. "California Casual," *Interiors*, January 1994, 80–1.

Weber, Matthew. "House of the Issue: Swatt | Mier," *The Architect's Newspaper*, September 29, 2011.

Weinstein, Elizabeth. "Animal Shelters Upgrade Creature Comforts," *Wall Steet Journal*, April 19, 2005, B1.

Western Red Cedar Lumber Association. "Architectural Design Awards 2008," *Architectural Record*. June 2009, 97, 101.

White, Debra J. "The Greening of Animal Shelters." Bark, November/December 2008, 34–8.

Digital Media

"A home in California designed for an art collector." *contemporist*, December 18, 2015.

"ARA Residence," *Architect Magazine*, November 24, 2014.

"Bold composition," *myTrends Home*, November 22, 2015.

"California home by Swatt Miers Architects face out over San Francisco Bay," *Dezeen*, July 23, 2015.

CED News. "House of the Issue; Swatt Miers—September 29 Architects Newspaper," *University of California, Berkeley, College of Environmental Design News,* Fall 2011.

CED News. "Tea Houses by Swatt | Miers Architects win Numerous Awards," *University of California, Berkeley, College of Environmental Design News*, Spring 2010.

"Couple create hillside haven at Sunol's Sinbad Creek," *SF Chronicle Online*, March 28, 2014.

Curwen, Rosell. "San Francisco AIA Design Winners Announced," *LA at Home, Los Angeles Times Blogs*, June 11, 2010.

"Escadaria dispensa ar-condicionado para refrescar casa na California," *Folha de Sao Paolo*, November 2, 2014.

Ferri, John. "1223 Upper Happy Valley Road Boasts Spectacular Vistas," *San Francisco Chronicle Online: Real Estate*, February 5 2012, K4–5.

"Garay Residence: A stunning contemporary home with spectacular views of San Francisco Bay," *10 Stunning Homes*, January 10, 2016.

"Glimpse Inside this Sprawling Home in Sunol," *SStyle Magazine*, October 16, 2015.

Henry, Christopher. "Garay Houses / Swatt Miers Architects," *ArchDaily*, August 10, 2011.

"Hogar artístico," *Architectural Digest Mexico*, November 23, 2015.

"House of the week: A Californian retreat," *World Architecture News*, July 24, 2015.

"Incredible Retrospect Vineyards Home," *Home Stratosphere*, February 2016.

Jordan, Gary. "Palm Springs Mid-Century Modernist Animal Shelter," March 24, 2012.

Keates, Nancy. "Modern Among the McMansions," *Wall Street Journal Online*, January 21, 2011.

King, Victoria. "OZ House / Swatt Miers Architects," *ArchDaily*, January 21, 2012.

King, Victoria. "Palm Springs Animal Care Facility / Swatt Miers Architects," A*rchDaily*, May 24, 2012.

King, Victoria. "Tea Houses / Swatt Miers Architects," *ArchDaily*, January 18, 2012.

Lee, Lydia. "Inside Out," *Modern Luxury Silicon Valley*, March 1, 2016.

"Palm Springs Animal Shelter," *Architizer,* April 14, 2013.

"Retrospect Vineyards House / Swatt Miers Architects," *ArchDaily*, August 19, 2015.

"Retrospect Vineyards House / Swatt Miers," *Architecture lab*, October 5, 2015.

"Retrospect Vineyards House by Swatt Miers offers expansive views of California's Wine Country," *Dezeen*, August 7, 2015.

"Retrospect Vineyards house in Windsor, California by Swatt Miers Architects," *AECCafe*, July 21, 2015.

"Retrospect Vineyards House," *archello*, September 4, 2015.

"Retrospect Vineyards Residence," *Architectural Record*, April 17, 2015.

"Retrospect Vineyards: Modern Californian Home by Swatt Miers," *10 Stunning Homes*, July 16, 2015.

"Rezidencia a Golden Gate híd közelében," *design.hu*, September 12, 2015.

Said, Carolyn. "Prefabricated Homes Go Upscale," *San Francisco Chronicle Online*, July 15, 2011.

"Sinbad Creek Residence: A modern house in rural Sunol, California," *10 Stunning Homes*, October 1, 2015.

"Sinbad Creek," *archello.* October 9, 2015.

Singhal, Sumit. "Palm Springs Animal Care Facility by Swatt | Miers Architects,"*AECCafe ArchShowcase*, June 10, 2012.

Spencer, Ingrid. "House of the Month," *Architectural Record Online*, March 2007.

Spencer, Ingrid. "The Tiptoeing House," *Business Week Online,* March 2007.

"Stein Residence," *Architect Magazine*, August 7, 2013.

"Step Inside a Gloriously Rebuilt 1970s Hilltop Home," *Curbed*, July 23, 2015.

"Studio and Retreats: Tea Houses," *Architectural Record Online: Featured Houses*, July 2010.

Tam, Pui-Wing and Keats, Nancy. "Silicon Valley Reboots," *Wall Street Journal Online*, October 24, 2012.

"Tea House by Swatt | Miers Architects," *Metalocus*, September 4, 2015.

"Tea Houses by Swatt | Miers," *Dezeen*, August 8, 2012.

"Tea Houses," *Architect Magazine*, June 3, 2013.

Tilton, Sarah. "Berkeley House Features Great Views, Modern Style," *Wall Street Journal Online*, February 23, 2012.

"Vidalakis Residence / Swatt | Miers Architects," *ArchDaily*, December 28, 2015.

"Vidalakis Residence in Portola Valley," *e-architect*, December 26, 2015.

"Vidalakis Residence," *Archello*, February 3, 2016.

"Vineyards Residence by Swatt | Miers Architects," *Home Adored*, August 10, 2015.

Weber, Matthew. "House of the Issue: Swatt | Miers," *The Architect's Newspaper*, September 29, 2011.

Williamson, Caroline. "A Modern Home in Rural Sunol, California," *Design Milk*, May 28, 2015.

图片版权信息
IMAGE CREDITS

All photography, renderings, drawings and plans published in this book have been supplied courtesy of Swatt | Miers Architects unless otherwise stated below.

Front Cover
Russell Abraham (Vidalakis House)

Foreword
Russell Abraham

Houses
Robert Swatt, FAIA

Residential Beginnings
Russell Abraham 13, 15
Tim Griffith 14

01 OZ House
Tim Griffith 16–17, 20–1, 22, 24–5, 26, 27 (top)
Robert Swatt, FAIA 19, 27 (bottom)

02 Sinbad Creek Residence
Russell Abraham 28–9, 31, 32, 34, 35 (top), 36–7, 38, 39 (top)
Robert Swatt, FAIA 35 (bottom), 39 (bottom)

03 Simpatico Prefab Prototype
Russell Abraham 40–1, 42 (top), 43, 44, 45, 46–7
Robert Swatt, FAIA 42 (bottom)

04 ARA House
Russell Abraham 48–9, 51, 52–3, 54, 55, 56 (top), 57
Robert Swatt, FAIA 56 (bottom)

05 Stein House
Russell Abraham 58–9, 61, 62–3, 64 (top), 65, 66

06 Vidalakis House
Russell Abraham 68–9, 71, 72–3, 74, 76–7, 78, 79 (top), 80, 81
Miya Muraki, AIA 75 (top—model creator and photography)
Robert Swatt, FAIA 75 (bottom), 79 (bottom)

07 Retrospect Vineyards House
Russell Abraham 82–3, 86, 88, 90, 91, 92, 93, 94, 95 (bottom)
Marion Brenner 85, 95 (top)
Robert Swatt, FAIA 87

08 Rashid House
Russell Abraham 96–7, 99, 100, 101, 102
Robert Swatt, FAIA 103

09 Oak Knoll House
Russell Abraham 104–5, 107, 108 (top), 109, 110–11, 112, 113, 114, 115
Robert Swatt, FAIA 108 (bottom)

10 Cheng-Brier House
Russell Abraham 116–17, 119, 120, 121 (top; bottom—model photography), 122–3
Niraj Kapadia 121 (bottom left—model creator)
Robert Swatt, FAIA 121 (bottom right)

11 Kuenster-Miers House
Russell Abraham 126–7, 129, 130–1, 133

12 Cheng-Reinganum House
Russell Abraham 134–5, 137, 138, 139, 140, 141 (top; bottom right—model photography)
Misa Grannis 141 (bottom right—model creator)
Robert Swatt, FAIA 141 (bottom left)

13 Mora Estates
DeLeon Realty / Anthony Halawa Photography 142–3, 145, 146, 147 (top)
Robert Swatt, FAIA 147 (bottom)

Houses in Progress
Robert Swatt, FAIA 148–9

14 House in Northern India
Pete Austin 151, 153
Niraj Kapadia and Kimie Jurado 152 (bottom—model creators and photography)
Robert Swatt, FAIA 152 (top)

15 Casa Santiago
Russell Abraham 157 (bottom—model photography)
Pete Austin 155, 157
Robert Swatt, FAIA 156
Ronald Tallon 157 (bottom—model creator)

16 Helios House
Russell Abraham 160 (top; middle; bottom—model photography)
Pete Austin 159, 161
Robert Swatt, FAIA 158
Jodie Zhang 160 (top; middle; bottom—model creator)

17 Amara House
Russell Abraham 165 (model photography)
Pete Austin 163, 164
Miya Muraki, AIA 165 (model creator)

18 Whispering Stones House
Pete Austin 167, 163
Johanna Malmerberg 168 (bottom right—model creator and photography)
Robert Swatt, FAIA 166

19 Irie House
Russell Abraham 170 (model photography), 172 (model photography)
Pete Austin 171, 173
Ronald Tallon 170 (model creator), 172 (model creator)

20 Orciuoli House
Russell Abraham 177 (bottom—model photography)
Pete Austin 175, 177 (top)
Misa Grannis 177 (bottom—model creator)

21 Westwind House
Russell Abraham 181 (top; middle; bottom—model photography)
Pete Austin 179, 180
Fanzheng Dong 181 (top; middle; bottom—model creator)
Robert Swatt, FAIA 178

22 Chalon Road House
Pete Austin 183, 185

23 Lee House
Russell Abraham 189 (top; middle; bottom—model photography)
Pete Austin 187, 188
Fanzheng Dong 189 (top; middle; bottom—model creator)
Robert Swatt, FAIA 186

24 River House
Russell Abraham 193 (top; middle; bottom—model photography)
Pete Austin 191, 192
Betty Nip 193 (top; middle; bottom—model creator)
Robert Swatt, FAIA 190

25 Black Point House
Russell Abraham 197 (model photography)
Pete Austin 195, 196
Betty Nip 197 (model creator)

26 Kailua Beach Houses
Russell Abraham 201 (model photography)
Pete Austin 199, 200, 201 (top)
Missa Grannis and Julie Liberman 201 (model creator)

27 SANA Residence
Pete Austin 205
Robert Swatt, FAIA 203, 204 (top; middle; bottom)

Commercial, Educational, & Institutional Projects
Markus Lui 204–5

28 Palm Springs Animal Shelter
Mark Davidson 212–13, 215, 217, 218, 219

29 Broadway Tennis Center
Russell Abraham 220–1, 223, 224, 225, 226

30 Annenberg Projects
Markus Lui 236, 237
Robert Swatt, FAIA 235
Erik Waterman 228–9, 231, 232, 234

Current & Recent Staff
Russell Abraham 240, 241

Biographies
Russell Abraham 242, 243

Back Cover
Marion Brenner (Retrospect Vineyards House)

致谢
ACKNOWLEDGEMENTS

The designs shown in this volume cover a prolific decade 2006–2016 that saw Swatt | Miers work expand into different regions of California, including Central Valley, Wine Country, and Los Angeles; different states, including Hawaii and Colorado; and other regions of the world, including Canada, India, and Spain. While each design is a unique solution for a particular site and for particular clients, there are threads that run through every project.

We are fortunate and grateful that that all of our clients for these projects came to share our vision of achieving meaningful modern architecture—architecture of our time. Certainly none of this work would have been possible without this shared vision and the courage to create something brand new—something that never existed before.

Every project in this book is the result of the efforts of many people. We are particularly appreciative of our talented professional staff who, with a commitment to making each project the best it can be, have made significant contributions to this work.

Similarly, our consulting engineers, contractors, and craftsmen have all contributed to turning these visions into architectural reality.

索引
INDEX

Amara House 162–5
Annenberg Projects 228–37
ARA House 48–57
Black Point House 194–7
Broadway Tennis Center 220–7
Casa Santiago 154–7
Chalon Road House 182–7
Cheng-Brier House 116–25
Cheng-Reinganum House 134–41
Helios House 158–61
House in Northern India 150–3
Irie House 170–3
Kailua Beach Houses 198–201
Kuenster-Miers House 126–35
Lee House 186–9
Mora Estates 142–7
Oak Knoll House 104–15
Orciuoli House 174–9
OZ House 16–27
Palm Springs Animal Shelter 212–19
Rashid House 96–103
Retrospect Vineyards House 82–95
River House 190–3
SANA Residence 202–5
Simpatico Prefab Prototype 40–7
Sinbad Creek Residence 28–39
Stein House 58–67
Vidalakis House 68–81
Westwind House 178–81
Whispering Stones House 166–9

Every effort has been made to trace the original source of copyright material contained in this book. The publishers would be pleased to hear from copyright holders to rectify any errors or omissions.

The information and illustrations in this publication have been prepared and supplied by Swatt | Miers Architects. While all reasonable efforts have been made to ensure accuracy, the publishers do not, under any circumstances, accept responsibility for errors, omissions and representations express or implied.